Affinity Chromatography

CHROMATOGRAPHIC SCIENCE

A Series of Monographs

Editor: JACK CAZES
Fairfield, Connecticut

Volume 15: Applications of Glass Capillary Gas Chromatography
Edited by Walter G. Jennings

Volume 16: Steroid Analysis by HPLC: Recent Applications
Edited by Marie P. Kautsky

Volume 17: Thin-Layer Chromatography: Techniques and Applications
Bernard Fried and Joseph Sherma

Volume 18: Biological/Biomedical Applications of Liquid Chromatography III
Edited by Gerald L. Hawk

Volume 19: Liquid Chromatography of Polymers and Related Materials III
Edited by Jack Cazes

Volume 20: Biological/Biomedical Applications of Liquid Chromatography IV
Edited by Gerald L. Hawk

Volume 21: Chromatographic Separation and Extraction with Foamed Plastics and Rubbers
G. J. Moody and J. D. R. Thomas

Volume 22: Analytical Pyrolysis: A Comprehensive Guide
William J. Irwin

Volume 23: Liquid Chromatography Detectors
Edited by Thomas M. Vickrey

Volume 24: High-Performance Liquid Chromatography in Forensic Chemistry
Edited by Ira S. Lurie and John D. Wittwer, Jr.

Volume 25: Steric Exclusion Liquid Chromatography of Polymers
Edited by Josef Janča

Volume 26: HPLC Analysis of Biological Compounds: A Laboratory Guide
William S. Hancock and James T. Sparrow

Volume 27: Affinity Chromatography: Template Chromatography of Nucleic Acids and Proteins
Herbert Schott

Other Volumes in Preparation

Affinity Chromatography

TEMPLATE CHROMATOGRAPHY OF NUCLEIC ACIDS AND PROTEINS

Herbert Schott

Universität Tübingen
Institut für Organische Chemie
Tübingen, Federal Republic of Germany

MARCEL DEKKER, INC. New York and Basel

Library of Congress Cataloging in Publication Data

Schott, Herbert, [date]
Affinity chromatography.

(Chromatographic science ; v. 27)
Bibliography: p.
Includes index.
1. Affinity chromatography. 2. Nucleic acids--Separation. 3. Proteins--Separation. I. Title. II. Series. [DNLM: 1. Chromatography, Affinity. 2. Nucleic Acids--isolation & purification. 3. Proteins--isolation & purification. W1 CH943 v.27 / QU 25 S375a]
QP519.9.A35S34 1984 574.19'245 84-17461
ISBN 0-8247-7111-7

MARCEL DEKKER, INC.
270 Madison Avenue, New York, New York 10016

Current printing (last digit):
10 9 8 7 6 5 4 3 2 1

PRINTED IN THE UNITED STATES OF AMERICA

Preface

Affinity chromatography, which uses the specific interactions between natural products for their separation, purification, and isolation, has provided an essential contribution to the rapid development of molecular biology and, especially, genetic engineering. The huge number of publications, especially in these current fields, ranges from the isolation of nucleic acids, genes, oligonucleotides, and nucleic acid fragments to the purification of special proteins and enzymes as well as investigations of protein-nucleic acid interactions. Since this interesting field has not been previously summarized sufficiently in books on affinity chromatography, it seems the right time to describe the published material more extensively in a book. Since the template functions of nucleic acids, specifically the base-pairing mechanism, are used for chromatography in almost all papers, so "Template Chromatography of Nucleic Acids and Proteins" seemed to be the obvious subtitle for the book.

The book is arranged as follows. First, various techniques are shown for immobilization of nucleic acid fragments, polynucleotides, and nucleic acids by which the desired separation of materials can be achieved. The main part of the book summarizes examples of applications. The isolation of DNA and its fragments, as well as RNA, is shown in various examples, with ample treatment of the isolation of mRNA. A chapter on protein isolations is followed by a chapter on enzymic synthesis or degradation of polynucleotides. The book ends with studies of peptide nucleotide interactions.

In the given context it is impossible to describe all published results. It is the purpose of this book to inform the reader about the practical methods, possibilities, and limits of template chromatography as well as to lead graduate students, young scientists, and those beginning to interest themselves in molecular biological questions to the original papers.

Herbert Schott

Contents

Abbreviations

A	adenosine
Ade	adenine
ADP	adenosine-5'-diphosphate
AE	aminoethyl
AMP	adenosine-5'-monophosphate
ATP	adenosine-5'-triphosphate
bp	base pairs
br^8ATP	8-bromoadenosine-5'-triphosphate
C	cytidine
cDNA	complementary deoxyribonucleic acid
cl^6ITP	6-chloroinosine-5'-triphosphate
CM	carboxymethyl
CMP	cytidine-5'-monophosphate
CNBr	cyanogenbromide-activated
CTP	cytidine-5'-triphosphate
Cyt	cytosine
dA	deoxyadenosine
DBM	diazobenzyloxymethyl
dC	deoxycytidine
DEAE	diethylaminoethyl
DMSO	dimethyl sulfoxide
DNA	deoxyribonucleic acid
DNase	deoxyribonuclease
ds	double-stranded
dT	deoxythymidine

EDTA	ethylenediaminetetraacetic acid
G	guanosine
GMP	guanosine-5'-monophosphate
GTP	guanosine-5'-triphosphate
Gua	guanine
H	heavy
HnRNA	heterogeneous nuclear ribonucleic acid
I	inosine
IMP	inosine-5'-monophosphate
kb	kilobases
L	light
mol wt	molecular weight
mRNA	messenger ribonucleic acid
mt	mitochondrial
mt DNA	mitochondrial DNA
N terminus	amino terminus
oligo(dT)	oligodeoxythymidylate
P-cellulose	phosphocellulose
poly(A)	polyadenylate
poly(C)	polycytidylate
poly(G)	polyguanylate
poly(I)	polyinosylate
poly(U)	polyuridylate
PVAL	poly(vinyl alcohol)
RNA	ribonucleic acid
RNase	ribonuclease
RNP	ribonucleoprotein
rRNA	ribosomal ribonucleic acid
SDS	sodium dodecyl sulfate
s^6ITP	6-thioinosine-5'-triphosphate
TCA	trichloroacetic acid
Thy	thymine
T_m	melting temperature
tRNA	transfer ribonucleic acid
U	uridine
UDP	uridine-5'-diphosphate
UMP = pU	uridine-5'-monophosphate
UTP	uridine-5'-triphosphate
UV	ultraviolet

1

Introduction

It is known that the exact replication of genetic information is controlled by the base pair mechanism described by Watson and Crick. This mechanism depends on the fact that Ade and Cyt undergo specific interaction with Thy and Gua, respectively.

The base pairs Thy·Ade and Cyt·Gua are thus referred to as complementary base pairs. Single strands of nucleic acids posessing complementary base sequences will aggregate to double strands under certain conditions. The association is not restricted to complementary single strands of nucleic acids but can, under certain conditions, also be observed among poly-, oligo-, and mononucleotides; nucleosides, and nucleobases.

Since base pairing is specific and reversible, it recommends itself as the basis of a form of affinity chromatography suitable for the separation and isolation of nucleic acids and their fragments. This is accomplished in practice by immobilizing defined nucleic acids, polynucleotides, oligonucleotides, or nucleic acid residues on a stationary phase. If the conditions of base pairing are fulfilled, the complementary partner in the mobile phase is strongly adsorbed by the stationary complementary partner, noncomplementary partners being retarded to only a small extent or not at all. After the elution of the noncomplementary components, the conditions of elution are altered such that base pairing is eliminated, and the adsorbed compounds are desorbed.

These immobilized compounds have the same template function as DNA, since only the nucleotide sequence of the immobilized compound determines which partner is complementary and which is thus adsorbed. The template function can be used additionally for the isolation of enzymes and other natural substances if they are capable of undergoing specific interactions with stationary nucleotide sequences.

This special form of affinity chromatography, which has shown rapid development in the last few years, may thus be termed *template chromatography*. The method has established itself as a useful technique which has recently received diverse applications in molecular biology as well as in the study of enzymes. The use of template chromatography in the study of enzymes is not surprising since about 30% of the approximately 2000 enzymes found in the cell participate in reactions involving nucleotides/nucleotide coenzymes.

The requirements of a suitable system for template chromatography are dictated by the moiety to be purified, the coupling of ligand to matrix, and the working milieu. For many research purposes details of molecular parameters may be vital: knowledge and reproducibility of coupling modes, matrix characteristics, and ligand-matrix-medium interactions then assume paramount importance. For large-scale production in particular, mechanical stability, rapid flow rate, resistance to microbial attack, and re-use are important factors. It will often be necessary to space the ligand from its support. This may be achieved either by coupling the ligand to one end of an "arm," the other end of which is subsequently attached to the carrier, or by coupling it to an arm already modifying the matrix. This precaution will be superfluous if the ligand is bulky enough to make its binding site(s) available.

The basic problem in template chromatography, that of immobilizing the nucleic acids and their fragments on a suitable support in order to form the stationary phase, may be solved in several ways, as will be shown.

2

Immobilization of Nucleoside Phosphates, Nucleosides, and Nucleobases

It is the aim of this chapter to review the basic strategies of immobilizing nucleoside phosphates, nucleosides, and nucleobases to insoluble supports. The affinity between the covalently bound nucleic acid component and the partner of the mobile phase to be isolated is highly stereospecific. The interaction does not only lead to a complex formation. The subsequent displacement of the molecule from the complex will depend (a) on the three-dimensional orientation, (b) on the group and position on the ligand molecule through which it is immobilized to the support, (c) to some extent on the chemical method by which the column is made, and (d) on the spacer distance between the ligand and the supporting matrix. Moreover, the exact mechanism by which the immobilized ligand interacts with the partner of the mobile phase is very important for the design of the optimal matrix.

Columns with immobilized nucleic acid components were prepared particularly with a view to using them to investigate the wide field of enzymes that interact with nucleic acids and/or their components. The variety of these enzymes required template columns in which different positions of the nucleic acid components are bound to polymer support.

I. IMMOBILIZATION OF NUCLEOSIDE PHOSPHATES

Figure 1, for example, demonstrates the possible positions on AMP that can be functionalized for the immobilization to a polymer sup-

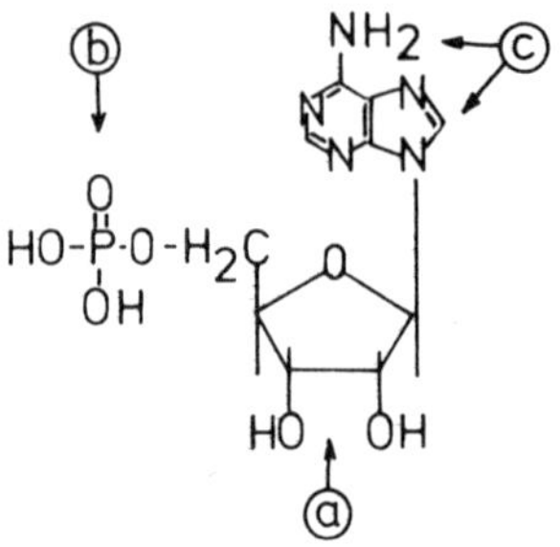

Figure 1 Positions of AMP that can be derivatized: (a) ribosyl-linked; (b) phosphate-linked; (c) N^6-linked and C-8-linked. (From Ref. 3.)

port. In essence, these can be classified into (a) phosphate-linked and (b) sugar-linked derivatives having an unsubstituted nucleobase, and (c) base-linked nucleoside phosphates containing free sugar and phosphate groups.

An immobilization of nucleotides in which several positions of the ligand are simultaneously involved, for example, amide linkage through an amino group, esterification of sugar hydroxyls, or anhydride formation with the phosphoric acid residue, results in an ill-defined matrix. Rather, it is better to aim for methods that result in a more defined matrix, the interaction of which can be more exactly predicted and controlled.

A lot of literature has been concerned with designing simplified chemical methods for preparing polymer support derivatized with nucleic acid components. The immobilization of ligands to several supports (e.g., CNBr-Sepharose, AE-Sepharose, CM-cellulose, AE-cellulose, and functionalized nylon) is described [1] and the yields of several reactions are summarized in Table 1.

The most exploited procedure for immobilization of nucleic acid components and their derivatives is the CNBr method, which has been reported by many laboratories. The results described, however, are very different [3-6]. The major problem in the preparation of adsorbents suitable for template chromatography is the design and synthesis of nucleotide ligands to be attached to the insoluble support.

Table 1 Methods of Linking Nucleoside Phosphates to Insoluble Supports

Ligand	Coupling yields μmol	Coupling yields %
Coupling of ligands to 500 mg of Sepharose by the CNBr method:		
p-Aminophenyl ATP		20
6-Aminohexanoyl ATP		90
s^6 ITP		43
Coupling of ligand to 500 mg of CM-cellulose by the mixed carbonic anhydride method:		
ATP	6.0	16
Coupling of halogen-substituted ligands to 200 mg of AE-cellulose:		
br^8 ATP	2.0	40
cl^6 ITP	24	72
Coupling of ligand to 500 mg of AE-Sepharose by the glutaraldehyde method:		
ATP	4.1	20
Coupling of ligands to 500 mg of AE-cellulose by bisimidate cross-linking:		
AMP	2.8	20
ATP	2.5	22
Coupling of ligand to 200 mg of nylon 6 powder by the methylation procedure:		
ATP	4.0	80

Source: Data from Ref. 1.

A. Phosphate-Linked Nucleoside Phosphates

Linkage through the phosphate groups is a method applicable to all nucleotides. The early approaches were based on the formation of phosphate esters (e.g., the synthesis of cellulose contain-

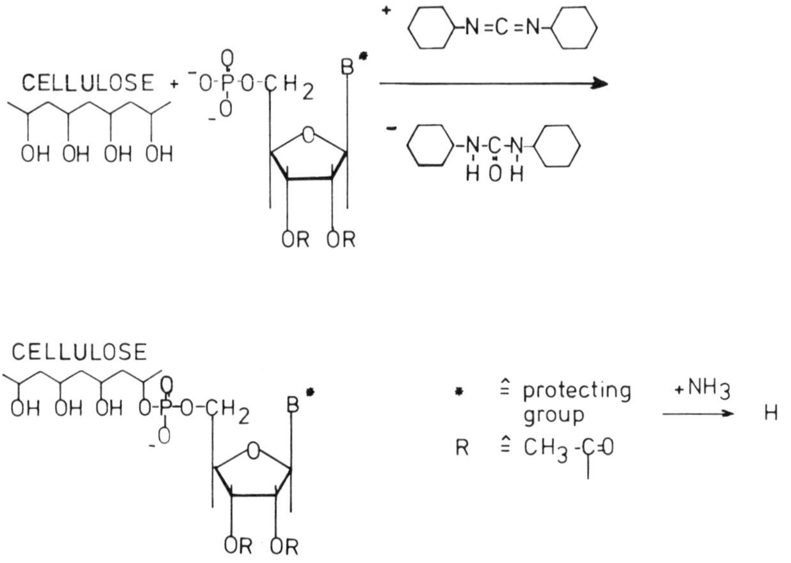

Scheme 1a

ing covalently bound nucleotides (Scheme 1a). For instance, thymidylate, adenylate, cytidylate, and guanylate cellulose are obtained by the condensation of the corresponding 5'-mononucleotides with hydroxyl groups of the cellulose support using dicyclohexylcarbodiimide in pyridine [7,8]. A similar carbodiimide reaction was used to link a 6-p-nitrobenzamido-hexan-1-ol spacer arm to the 5'-phosphate of 2'-deoxyuridylic acid, prior to attachment to agarose [9]. Immobilization through p-aminophenyl esters of the nucleotides has also been used [10]. Coincidental production of symmetrical pyrophosphate is a potential problem during the synthesis of these phosphate esters. Anion displacement [11] and imidazolide synthesis [12] are proposed to overcome this undesired side reaction. According to Scheme 1b, nucleoside phosphate can be derivatized to a ligand that contains one more phosphate group than the starting material [2]. The amino terminal of this nucleoside phosphate derivative can be used for attachment to CNBr-agarose. The synthesis of the P^1-(6-aminohex-1-yl)-P^2-(5'-nucleosyl)pyrophosphate derivative of uridine [13], adenosine [14], deoxyguanosine [15], and cytidine [16] was based on this procedure. Adsorbents containing uridine-5'-phos-

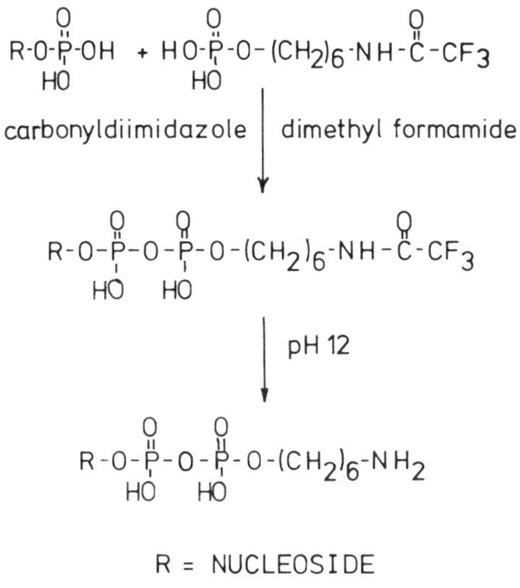

Scheme 1b

phate or pyrophosphate residues linked to Sepharose through phosphoamide bond were prepared through reaction of N^1-(uridine-5'-phosphoryl)- or N^1-(uridine-5'-pyrophosphoryl)-1,6-diaminohexanes with CNBr-activated Sepharose [34].

B. Sugar-Linked Nucleoside Phosphates

Linkage through the sugar hydroxyl groups is the easiest way of immobilizing nucleotides. Initially, carbodiimide-promoted condensation reactions were used [17], but this technique has been superseded by the hydrazone procedure, which involves the reaction of specifically modified groups of the ribonucleotides, and thus ensures a more defined chemical linkage (Scheme 2a). The 2'- and 3'-hydroxyl groups of ribonucleotides are oxidized to aldehyde functions, which are then linked to Sepharose [2,18]- or agarose [19]-bound hydrazide. The immobilization is carried out under mild conditions, the reaction is fast, and the binding capacity is very high. This method has been employed extensively and has proved to be the most useful of the ribosyl linkages. Table 2 shows some of the nucleotides immobilized by this method.

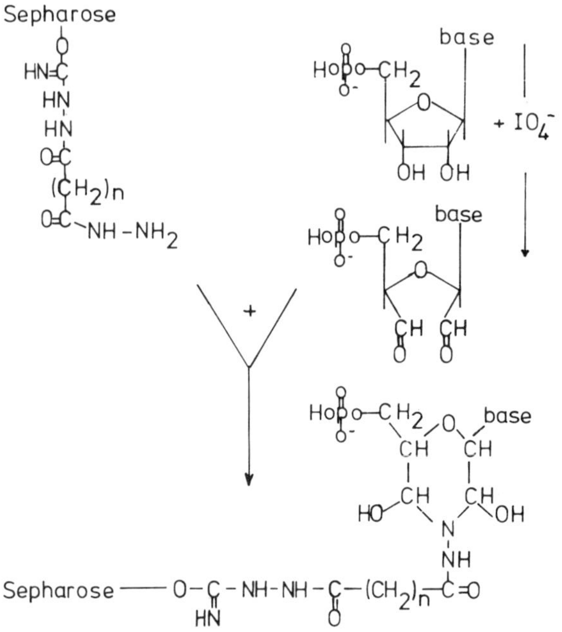

Scheme 2a

The same procedure can also be used for coupling to polyacrylic hydrazide support [20]. Alternatively, oxidized nucleotides can be linked first with a dihydrazide, followed by coupling to CNBr-agarose [18,21,22]. The aldehyde groups of the oxidized ribonucleoside phosphates can also be coupled with AE-cellulose [8] or 6-aminohexyl-agarose/Sepharose [23] by the formation of a Schiff's base which is subsequently reduced to a stable alkylamino linkage.

The aminooxybutylcellulose (AB-cellulose) application is reported for immobilizing periodate-oxidized nucleosides; their mono-, di-, and triphosphates; as well as tRNA, poly(A), poly(U), and S-adenoxyl-L-homocysteine [569]. Other possibilities for nucleotide immobilization through saccharide residues were exemplified with isopropylidene-5'-deoxy-5'-oxadenosine and ribose 5'-phosphate. At pH 5 nonoxidized cytidine derivatives and cytidine-containing polynucleotides were shown to react easily with AB-cellulose via cytidine moieties, whereas immobilization of

Table 2 Coupling of Nucleoside Phosphates to Sepharose or Agarose Hydrazide

	Nucleoside phosphates (μmol)		
	Per ml of Sepharose		Per ml of agarose
Ligand	Added	Bound	Bound
AMP	4.4	4.0	3.4
ADP	4.8	4.3	8.9
ATP	7.0	3.5	5.1
UMP	3.9	3.6	
UDP	–	–	4.0
UTP	4.0	3.2	
CMP	4.8	3.9	
CTP	4.2	3.5	
GMP	3.7	2.9	
GTP	4.5	3.3	

Source: Data from Refs. 18 and 19.

poly(A) and poly(U) proceeded at a considerably lower rate. In order to attain selective immobilization, AB-cellulose was subjected to modification with α-ketoglutaric acid, which resulted in an enhanced rate for immobilization of positively charged components.

A convenient alternative method to periodate oxidation is the introduction of ketone derivatives such as ethyl levulinate as protecting groups for the 2',3'-vicinal hydroxyl system of ribonucleosides (Scheme 2b). The derivatization of IMP according to this concept is carried out by producing inosine 2',3'-O-(ethyl-levulinate)acetal, phosphorylating to the corresponding IMP analog and finally hydrolysis to the levulinic acid acetal and condensation of the terminal carboxylate with 1,6-diaminohexane in a carbodiimide-activated reaction [24,25]. The approach yields an inosine phosphate derivative suitable for direct attachment to CNBr-agarose. Both the periodate oxidation procedure [21] and the procedure reported in [26] yield analogs which are coupled to pre-

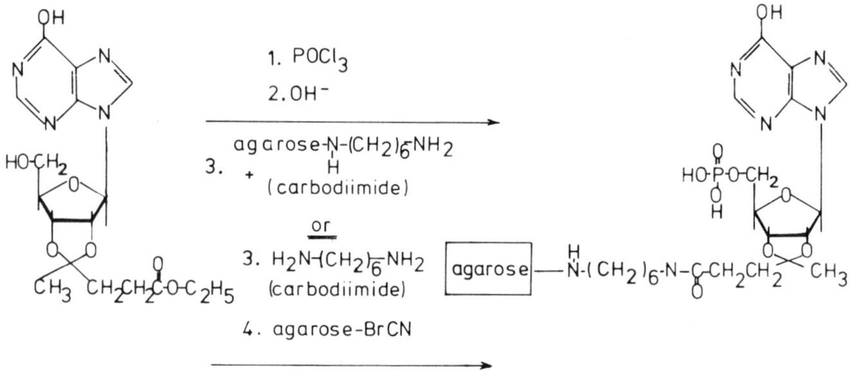

Scheme 2b

formed spacer arm matrices and thus are more likely to encounter problems of nonspecific adsorption to supernumerary spacer molecules in the adsorbent [27].

C. Nucleobase-Linked Nucleoside Phosphates

Derivatives of nucleoside phosphates containing a primary amino group can be linked easily to CNBr-agarose and CNBr-Sepharose [1,3]. Most efforts have been directed toward linking adenosine nucleotides through the adenine ring to an agarose support (Scheme 3). Techniques for the immobilization of nucleoside phosphates through the 6- or 8-position of the purine or through the 5-position of their pyrimidine bases have been reported in detail [2,3]. N^6-(6-aminohexyl)-ATP and 8-(6-aminohexyl)amino-ATP coupled to agarose are commercially available.

Adenine nucleotides have also been immobilized on modified Sepharose 4B or Dextran T40 with glutaraldehyde and reduced with KBH_4. Binding was dependent on pH and the nature of the amino group on the modified polysaccharide [35]. This method is suitable for the coupling of adenine nucleotides via the primary amino group to 1,4-diaminobenzene derivatives of polysaccharides. The results show that the method has the advantages of simplicity, speed, and good yields. The procedure could be applied to the immobilization of other compounds containing primary amino groups.

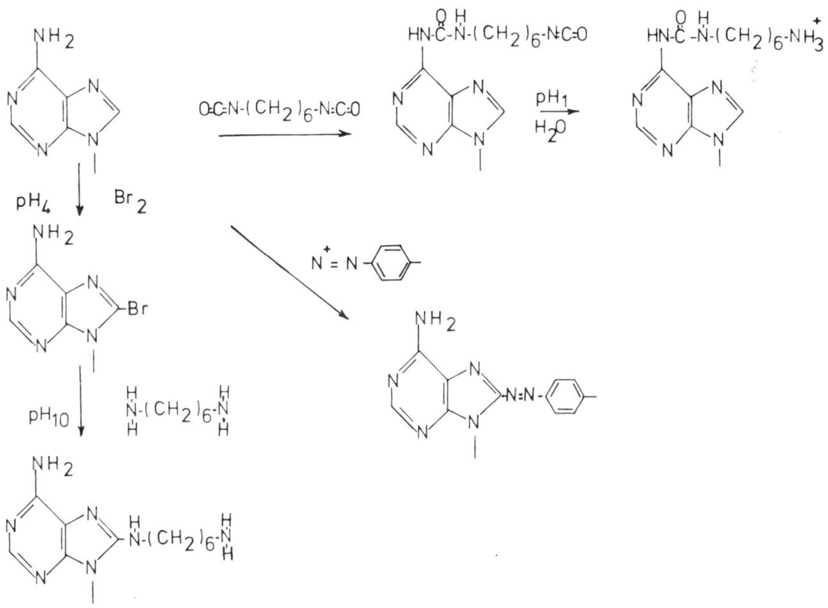

Scheme 3

A simple one-step reaction for preparing N^4-aminoalkylcytidine derivatives suitable for immobilization to an insoluble support has also been reported [28]. The synthesis and characterization of 8-(6-aminohexyl)inosine-5'-monophosphate bearing a spacer molecule with terminal amino group suitable for coupling to CNBr-agarose have been described [25].

Two guanine nucleotide analogs, 8-(6-aminohexyl)aminoguanosine-5'-triphosphate and 8-(6-aminohexyl)aminoguanosine-5'-diphosphate, were synthesized from the monophosphate by phosphorylation with pyrophosphate or orthophosphate [568]. Both guanosine nucleotides were used as ligands for affinity chromatography by attaching them to Sepharose 4B by the cyanogen bromide method.

II. IMMOBILIZATION OF NUCLEOSIDES AND NUCLEOBASES

A. Covalent Coupling of Nucleosides

The procedure described in Sec. I.B.3 can also be applied in general for the immobilization of nucleosides. Further methods for attachment nucleosides and their derivatives on CNBr-agarose, AE-cellulose, and functionalized nylon powder have been reported [1]. A series of cellulose derivatives has been synthesized by the reaction of epichlorohydrin and A, C, G, U, dT, and I [29]. Cellulose was chosen for the support because of its ability to withstand changes of salt concentration, pH, and hydrostatic pressure without gross changes in bed volume or flow rate.

B. Cross-Linking Copolymerization of Polymerizable Nucleoside Derivatives

The radical copolymerization of nucleosides whose hydroxyl groups are singly or multiply esterified with methacrylic acid yields insoluble cross-linked nucleoside gels [30-32] (Scheme 4).

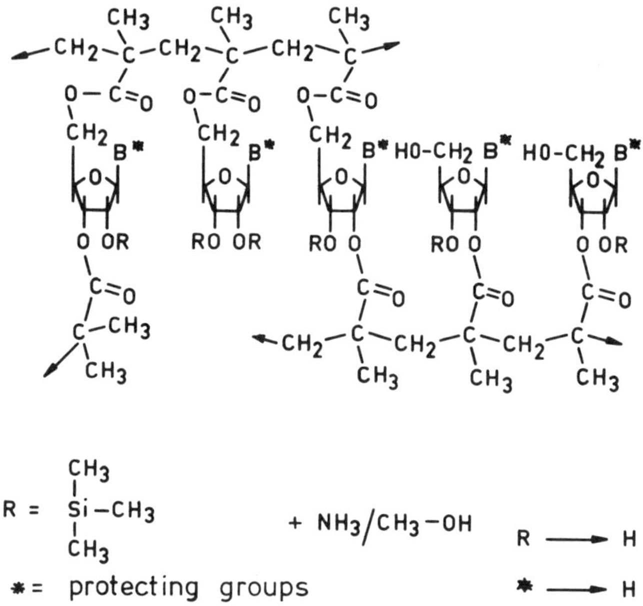

Scheme 4

These nucleoside gels possess very high concentrations of covalently bound nucleoside residues. Moreover, the gels may be used either in organic or aqueous medium and show good flow rates in column chromatography. They can be prepared rather easily on any desired scale.

C. Covalent Coupling of Nucleobases

In principle, the methods in Sec. I.C can also be used for immobilization of nucleobases to a polymer support. A guanine derivative of cellulose has been synthesized by linking guanine with p-diazobenzoate ester of cellulose [33]. The preparation of uracil-, thymine-, and cytosine-modified silica gel has been described [141] (Scheme 5).

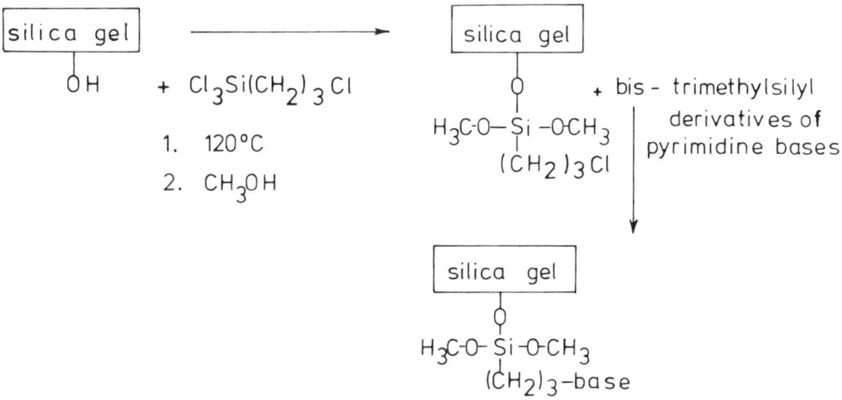

Scheme 5

3

Immobilization of Oligonucleotides

Several strategies have been developed to immobilize oligonucleotides on insoluble supports. In principle, three different approaches have been investigated: covalent linkage of the oligonucleotides via their terminals using (a) one-step or (b) multistep reaction and immobilization (c) via a secondary valence.

I. COVALENT BINDING OF THE OLIGONUCLEOTIDE TERMINALS

A. One-Step Reaction

In a well-investigated approach, mononucleotides are polycondensed chemically using dicyclohexylcarbodiimide in pyridine, and on the addition of cellulose, the activated terminal phosphate groups of the oligonucleotides formed are condensed with the cellulose hydroxyl groups (Scheme 6). The final product should thus consist of cellulose containing a mixture (in size) of the bound homopolymer chains (of up to about 10 nucleotides in length) connected by phosphodiester linkages at the 5' terminals or the 3' terminals of the chains depending on whether the 5' or the 3' nucleotides are used in the polycondensation reaction. Although this one-step method is an efficient one in that most of the regular polynucleotide chains in the reaction mixture are covalently bound, the polynucleotides that may be incorporated in this way are limited in base sequence and chain length to those which can be prepared chemically in anhydrous solution. According to this strategy, oligo(dT) is immobilized chemically by the

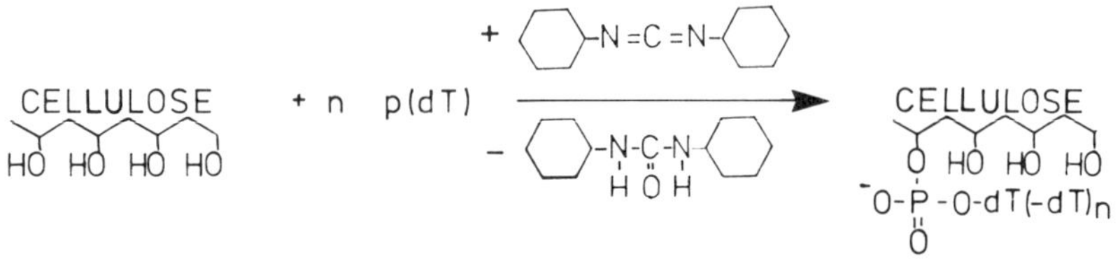

Scheme 6

reaction of thymidine 5'-phosphate with dicyclohexylcarbodiimide. In the case of polycondensation of deoxycytidylic, deoxyadenylic, and deoxyguanylic acid it is necessary to use protecting groups for the nitrogens of the nucleobases. The protecting groups are then removed after the linkage of the oligonucleotides to the cellulose. Using the one-step method, oligo(dT)- [7,8,36-38], oligo-(dC)- [8,37], and oligo(dA)-cellulose [8,37,38], and guanosine-5'-phosphoryl cellulose [8] have been synthesized.

B. Multistep Reaction

A more versatile, multistep procedure has been reported [39] which is applicable to polynucleotides of both synthetic and natural origin and has the added practical advantage that polymers may also be incorporated onto cellulose paper strips as well as cellulose powder. In this system the terminal phosphate group or polyphosphate group of a polynucleotide is activated in aqueous solution under mild conditions with a water-soluble reagent, and although the mechanism of the reaction differs from that operating in anhydrous synthesis, the product obtained is thought to be similar, as it consists of cellulose containing covalently bound polynucleotide chains each of which is connected to the cellulose at its 3' or 5' terminal by a phosphodiester linkage or a linkage involving a diester of polyphosphoric acid. According to this strategy, one end of an synthetic-defined oligonucleotide has been linked to cellulose paper in high yield by phosphate ester formation, using the water-soluble carbodiimide, 1-cyclohexyl-3-(2-morpholinoethyl) carbodiimide metho-p-toluenesulfonate [40]. The resultant oligodeoxyribonucleotide-cellulose papers were converted into a form suitable for column chromatography. A disadvantage of this method is the occurrence of side reactions at nucleobases and with internucleotide phosphodiesters.

Nucleotides and polynucleotides with terminal phosphoryl groups can be covalently linked in a carbodiimide reaction to water-soluble

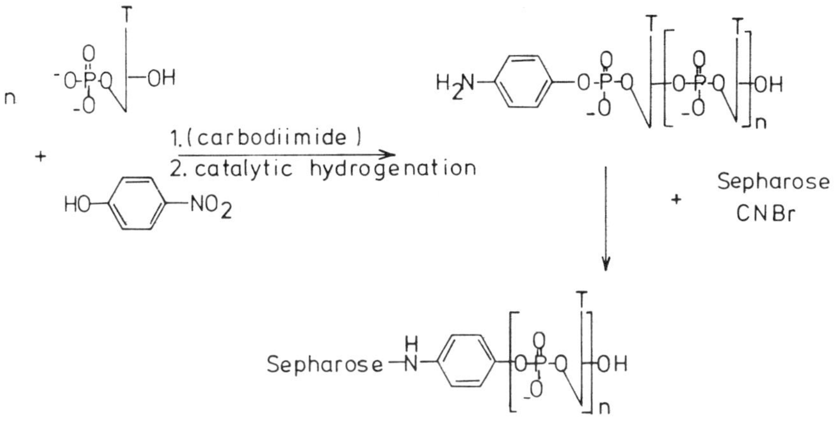

Scheme 7

polysaccharides such as Ficoll and dextran [417]. Small, dialyzable nucleotides can be converted to nondialyzable macromolecules of high molecular weight. Hybrid molecules of polynucleotides and polysaccharides have novel properties such as a reduced density in cesium chloride and modified end groups.

As an alternative to the immobilization of oligo(dT) via their terminal phosphate groups, the preparation of p-aminophenyloligo-(dT)-Sepharose has been reported [41] (Scheme 7). The attachment of 5'-NH_2-terminated oligo(dT) by coupling with CNBr-activated cellulose occurs in high yield and exclusively via the 5' terminal [42]. However, a problem that remained to be solved is the leakage of nucleotides from the solid support in basic medium and at elevated temperatures, a phenomenon inherent to CNBr activation. Mixed anhydride of oligonucleotides and mesitoic acid chloride were also applied to achieve immobilization of oligonucleotides via their phosphomonoester end groups on aminohexyl-Sepharose [43].

Methods are described by which primer molecules such as U-U and oligo(dT) can be coupled with high efficiency to an insoluble polymer, such as hydroxypropylated Sephadex G-50, by one covalent linkage [44]. In one procedure aliphatic dicarboxylic dichlorides (e.g., adipoyl dichloride) are used to serve as spacers of variable length and for anchoring the primer molecule U-U. The other method involves pU as an anchor for $(dT)_3$ and $(dT)_6$, which are coupled to the polymer using condensation reactions with 2,4,6-triisopropylphenylsulfonyl chloride. In both cases the homogeneous primer molecules are bound specifically to the polymer. The immobilized oligonucleotides can be cleaved from the polymer under conditions which are not damaging to ribo- and deoxyribopolynucleotides.

The immobilization of three different functionalized 5'-NH_2-terminated oligonucleotides (Fig. 2) on 2,4,6-trichloro-s-triazine- or 2-amino-4,6-dichloro-s-triazine-activated cellulose has been reported [45]. The immobilized oligonucleotides could be released quantitatively from the solid support by base or acid treatment or RNase digestion.

As an alternative to the polysaccharide matrix, cross-linked poly(vinyl alcohol) has been used as insoluble support for the attachment of oligo(dG) [46]. Protected homologs of deoxyriboguanylic acid are synthesized on a preparative scale by condensation of monomers according to the diester method. Oligomers with four and more monomeric units are isolated and bound to the free hydroxyl groups of an insoluble poly(vinyl alcohol) (PVAL) gel

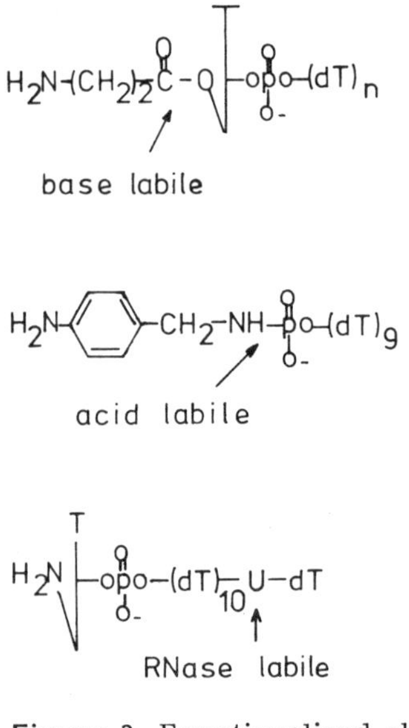

Figure 2 Functionalized oligonucleotides suitable for reversible attachment to activated cellulose. (Adapted from Ref. 45.)

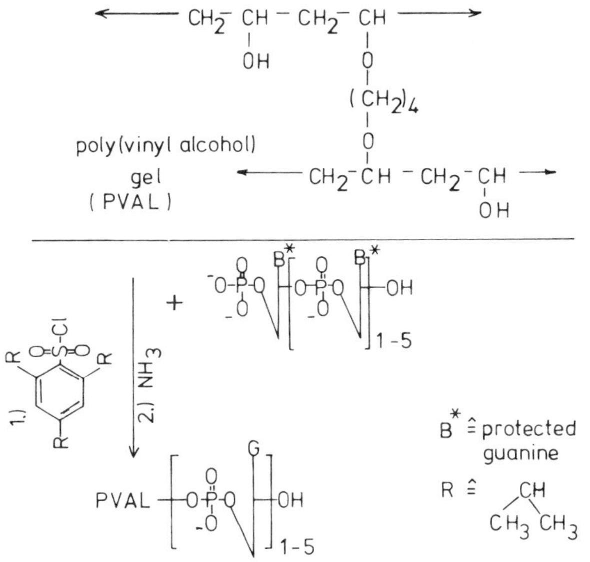

Scheme 8

via ester linkage with their 5'-phosphate groups (Scheme 8). The oligo(dG) gel, obtained after cleavage of the protecting groups, is suitable for preparative template chromatography of nucleotide mixtures.

II. SECONDARY VALENCE BINDING OF POLYMER-BOUND OLIGONUCLEOTIDES

The covalent attachment of oligonucleotides of known sequence and chain length to an insoluble support has the fundamental disadvantage that side reactions can occur during the chemical immobilization of the oligonucleotide. Also, the methods described for the covalent linkage of oligonucleotides have been developed for an analytical scale; preparative scale-up appears to be impractical.

The following strategy avoids the disadvantages of covalent immobilization. According to this method, defined oligonucleotides are covalently bound to the free hydroxyl groups of a soluble poly(vinyl alcohol) by means of their 5'-phosphate terminals.

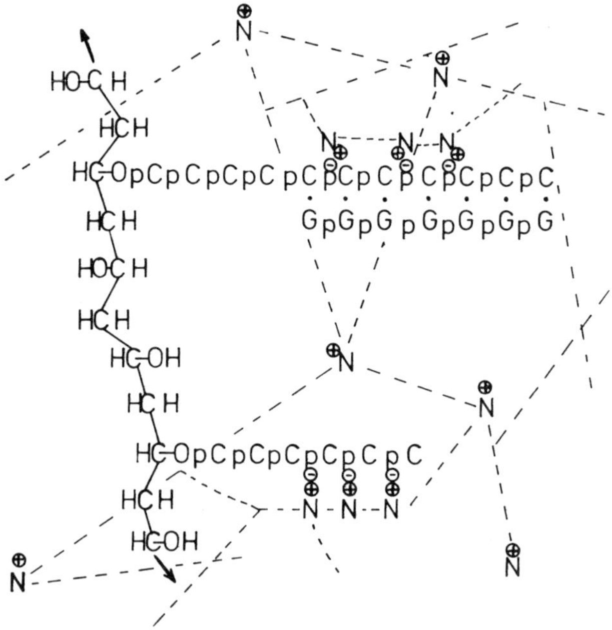

Figure 3 Schematical structure of the PVAL-p(dC)$_n$-DEAE-cellulose that adsorbed oligo(dG) by base pairing.

The polymer-bound oligonucleotide thus obtained can be purified and is subsequently irreversibly adsorbed as polyanion on the anion exchanger DEAE-cellulose using a batch procedure. The method allows the preparation of DEAE-cellulose loaded with oligonucleotides of known constitution on any scale required and avoids undesired side reactions during the immobilization. The oligonucleotide-DEAE-cellulose possesses both high capacity and good flow rate, under low pressure, when used as column packing. The schematical structure of PVAL-p(dC)$_n$-DEAE-cellulose is given as an example for the other oligonucleotide-DEAE-cellulose and is demonstrated in Fig. 3.

PVAL-p(dC)$_n$-DEAE-cellulose [47] has been synthesized by the following reaction steps. N-Anisoyldeoxyribocytidine-5'-phosphate is chemically polycondensed. From the polycondensate, oligomers containing four or more monomeric units are isolated chromatographically and subsequently linked to a soluble poly-

(vinyl alcohol). After cleavage of the nucleobase-protecting groups, the unreacted oligo(dC) is separated from poly(vinyl alcohol)-bound oligomers through ultrafiltration. DEAE-cellulose is subsequently coated with this polymer-bound oligo(dC). PVAL-p$(dT)_n$- [48] and PVAL-p$(dA)_n$-DEAE-cellulose [49] can be synthesized in an analogous manner.

5'-Tritylated oligonucleotides binding hydrophobically to low substituted trityl cellulose/Sepharose retain their hydrogen-bonding specificities for complementary sequences. This method, constituting a mode of attaching ligands to solid supports, proceeds with 100% yield. The salt, dielectric constant, and temperature dependence of these noncovalently anchored ligands permits the isolation of a variety of RNA. Medium trityl-Sepharose has a high binding specificity for poly(A) and poly(A) mRNA, equivalent to oligo(dT)-cellulose [301].

4

Immobilization of Polynucleotides and Nucleic Acids

Since in template chromatography interactions between the components should mimic the conditions of solution, careful consideration has to be given to the nature of the insoluble support, the ligand, and the method of immobilization. Ideally, the nucleic acids should be irreversibly immobilized and retain their native structure. In view of the very different problems that have to be solved using template chromatography, several more or less suitable strategies for the immobilization of nucleic acids and polynucleotides have been developed. The most significant approaches are reviewed [50-52] and summarized in this chapter and Table 3.

I. IMMOBILIZATION VIA PHYSICAL AND MECHANICAL FORCES OR IRRADIATION

A. Adsorption onto Cellulose

The very simple attachment of single- and double-stranded DNA to cellulose by drying and subsequently lyophilization of a DNA-cellulose paste leads to fairly stable DNA-cellulose complexes [53-55]. This is probably the most convenient method to use in preparing DNA-cellulose, but the DNA is bound reversibly and is slowly desorbed. No DNA is bound to the cellulose without drying. Moreover, if the air-drying step is omitted and a damp paste is lyophilized instead, the fraction of DNA bound to the cellulose decreases by about half. Single-stranded DNA-cellulose must be

Table 3 Immobilization Procedures of Nucleic Acids and Polynucleotides

Method	Ligand	Support	References
I.A Adsorption	(1) Native DNA	Cellulose	53-55
	(2) High-molecular-weight denatured DNA	Cellulose	53-55
	(3) Single-stranded mammalian DNA	Cellulose	53-55
	(4) R_{17} RNA	Cellulose	53-55
I.B Ultraviolet irradiation	(1) Poly(U)	Poly (vinyl chloride) beads	57
	(2) Poly(U)	Fiberglass	58
	(3) Poly(C)	Fiberglass	58
	(4) Poly(A), poly(C), poly(U)	Cellulose	59, 60
	(5) Native DNA	Cellulose	61
	(6) f2 single-stranded RNA	Cellulose	62
	(7) rRNA	Methylated cellulose	63
I.C Entrapment	(1) Single-stranded DNA	Cellulose acetate	64
		Agar	64, 65
	(2) fd DNA, calf thymus DNA denatured	Agar	66
	(3) Circular double-stranded DNA	Agar	67
	(4) Denatured DNA	Nitrocellulose	68, 70
	(5) Restriction fragments of DNA	Nitrocellulose	69
	(6) Poly(A), poly(G), poly(I), poly(C)	Agarose	71
	(7) Poly(U)	Polyacrylamide beads	72
	Salmon sperm DNA, native DNA	Polyacrylamide beads	73
II.A CnBr activation	See Table 4	Agarose/Sepharose	74-83

II.B	Azo coupling	(1)	Partially single-stranded SV40 DNA	Cellulose	84
		(2)	Sheared single-stranded DNA	Aminopentane Sepharose	85
		(3)	Single-stranded DNA, RNA	Diazobenzyloxymethyl paper, cellulose	86-90
		(4)	Double-stranded DNA	Enzacryl gel	91
II.C	Carbodiimide and carbonyldiimidazole activation	(1)	T_4-DNA, RNA, polynucleotides	Phosphocellulose	92, 93, 96, 97
		(2)	Sonicated rat liver DNA	Sephadex G-200	94
		(3)	Sheared single-stranded DNA	Aminopentane Sepharose	85
		(4)	DNA fragments	Silica carrier	95
II.D	Cyanuric chloride activation	(1)	DNA	Cellulose	
				Powder	98
				Paper	99
II.E	Condensation with CM-cellulose	(1)	Polynucleotide, DNA, RNA	CM-cellulose	1, 100
II.F	Bisoxirane activation	(1)	Single-stranded calf thymus DNA, tRNA	Sepharose, cellulose	101
		(2)	Cloned double-stranded cDNA	Cellulose	102
III.A	Oxidation of RNA	(1)	tRNA, RNA oligonucleotide	AE-cellulose	8, 104, 105
		(2)	tRNA	Agar-polyacrylhydrazide	107
		(3)	mRNA	Agar-polyacrylhydrazide	108
		(4)	tRNA	Hydrazinyl-Sepharose	109
		(5)	RNA	Agarose	110
		(6)	tRNA	Diverse supports	111
		(7)	tRNA, rRNA	Hydrazinyl-agarose	112, 113
III.B	Complexation with dihydroxyboryl groups	(1)	Ribonucleotides, tRNA	Cellulose, polyacryl derivatives	115-127

made by a modified procedure. Complexes between high-molecular-weight denatured DNA and cellulose are quite stable, whereas RNA and sonicated native DNA complexes with cellulose break down more rapidly. However, RNA of bacteriophage R_{17} forms with cellulose complexes, which can be used for template chromatography of RNA-binding protein. Drying tRNA with cellulose does not produce any detectable complex. It appears that a weak adsorption occurs at several sites on a single nucleic acid molecule, and that the cumulative effect of many such interactions holds long nucleic acid to the cellulose more firmly than shorter molecules are held. Due to the mild procedure, damage to the ligand is kept to a minimum. Since DNA-cellulose prepared by adsorption exhibits only limited stability to desorbing conditions such as high temperatures, formamide, and low ionic strengths of the buffers used, it cannot be recommended for hybridization experiments. If irreversible binding of DNA to cellulose is desired, a number of procedures have been developed.

B. Ultraviolet Irradiation

Nucleic acids and polynucleotides can be immobilized to insoluble supports by irradiating with ultraviolet (UV) light as proposed by [56]. The UV irradiation is presumably a consequence of the production of intermolecular cross linkages via the nucleobases of the polynucleotides, and the supports may serve only to immobilize the cross-linked nucleic acids and polynucleotides. It is possible that no covalent linkages are formed between support and ligand. However, the exact mechanism of the immobilization is unknown and other explanations are also to be considered. The method has been used to immobilize polyribonucleotides to poly(vinyl chloride) beads and to fiberglass. Poly(U) has been attached to poly(vinyl chloride) beads [57]; both poly(U) and poly(C) have been linked in the presence of UV light to fiberglass filters [58]. The attachment of poly(A) [59,60], poly(U), and poly(C) to cellulose by UV irradiation has also been reported [60].

Using UV irradiation, native DNA has been immobilized to cellulose in apparently large amounts and high yields [61]. The resulting DNA-cellulose is stable even to strong desorbing conditions and elevated temperature. UV irradiation has been extended to the attachment of viral RNA to cellulose [60] and of rRNA to methylated cellulose [63]. In a methylated cellulose the carboxyl groups of the cellulose are blocked and unspecific adsorption is minimized.

C. Entrapment Within the Insoluble Support

Methods have been described for the immobilization of high-molecular-weight single-stranded DNA by trapping in gels of cellulose acetate or agar [64]. Significant application of the procedures requires appropriate attention to a number of precautions [65]. The entrapment of phage fd DNA and denatured calf thymus DNA in 4% agar during gel formation has been reported [66]. The DNA content of gel fragments is stable under the conventional conditions of enzyme purification. The relatively high amounts of protein that are retained by DNA-agar due to unspecific interactions with the matrix can be overcome by carrying out several cycles of adsorption and elution. The entrapment procedure also allows immobilization of circular double-stranded DNA [67].

Denatured DNA [68] and restriction fragments of DNA [69] immobilize on nitrocellulose filters. Conditions are also described for the immobilization of DNA in alkali to cellulose nitrate filters [70].

Poly(A), poly(G), and poly(I) associate reversibly with agarose gels at high NaCl molarities over the pH range 6-10 at 20-40°C. Pyrimidine polyribonucleotides poly(C) and poly(U) could not be immobilized in agarose gels under the foregoing conditions. However, poly(C) could be immobilized in agarose without precipitation between pH 3.2 and 4.0 [71]. Association of poly(G) and poly(I) with agarose appears to decrease progressively with deprotonation of their purine residues, and both polymers interact with the gel very weakly above pH 10 regardless of NaCl concentrations. The binding to agarose of these polymers at pH 7.5 is also strongly influenced by temperature in the range 20-40°C. The association of single-stranded poly(A) is shifted toward higher NaCl molarities only by increased pH; its binding is also little affected by temperature in the range 20-40°C. At NaCl molarities effecting saturation retention in agarose and at neutral pH, the immobilization of several polynucleotides could be prevented by urea in a concentration-dependent manner. The corresponding profiles of urea molarity appear to disclose a number of hydrophobic interactions between polynucleotides and agarose, some of which might be relatively strong, especially in the case of poly(A).

Acrylamide gel beads are used as a matrix to immobilize poly(U) or DNA [72,73]. The gel is made by dissolving DNA in the acrylamide solution before cross-linking polymerization. The resulting gel can be used in the column and is stable at elevated temperature and in hybridization experiments. The pore size of the gel particle is such that the DNA remains trapped, unless it is attacked enzymatically. So far only native DNA has been used in the gel, but there is no reason why denatured DNA, RNA, or in fact any other macromolecule should not be used since the pore size may be varied at will; therefore, even small molecules can be entrapped in the particles. If the molecules that one seeks to purify are too large to enter the gel particle, this procedure can still be used, for there appears to be enough immobilized DNA protruding from the surface to be able to interact efficiently with the molecules in the liquid phase.

II. IMMOBILIZATION USING CHEMICAL ACTIVATION

A. Condensation with CNBr-Activated Polysaccharides

The activation of polysaccharide by CNBr is quite easy and enables the covalent linkage of a wide spectrum of compounds to insoluble supports (Scheme 9; see also Table 4).

Single-stranded DNA, RNA, and polyribonucleotides may be efficiently bound to CNBr-Sepharose [74-76,78]. Double-stranded DNA in which the nucleobases are buried on the inside of the helix binds very inefficiently to CNBr-Sepharose. In contrast,

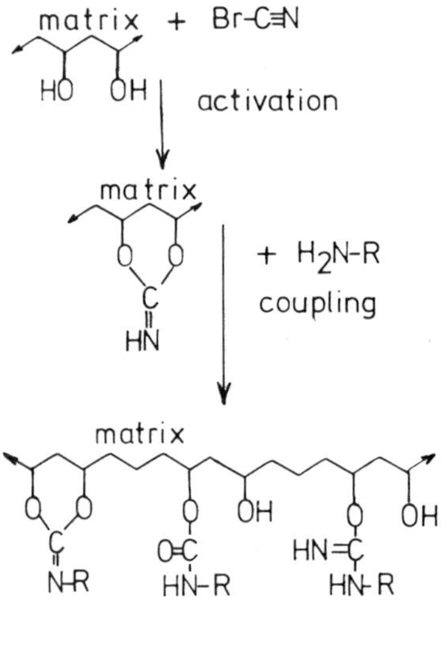

Scheme 9

Table 4 Results of Coupling Polynucleotides and Nucleic Acids by the CNBr Method

Polynucleotide or nucleic acids	Sepharose	pH; buffer	Percent coupled
Poly(I)	500 mg	6.0; collidine (I = 0.06)	42.0
	500 mg	8.0; collidine (I = 0.06)	Less
Poly(C)	500 mg	6.0; collidine (I = 0.06)	20.0
Poly(A)	10 ml	8.0; phosphate (0.05 M)	98.0
[^{14}C]Poly(dA-T)	5 ml	8.0; phosphate (0.05 M)	75.0
tRNA	500 mg	9.0; HCO_3^-	15.0
DNA Calf thymus	500 mg	9.0; HCO_3^- (0.08 M NaCl)	15.0
Herring sperm (denatured)	500 mg	9.0; HCO_3^- (0.08 M NaCl)	25.0
[^{3}H] DNA *E. coli* (native)	5 ml	10.65; HCO_3^- (0.05 M)	2.0
E. coli (denatured)	5 ml	8.0; phosphate (0.05 M)	20.9
[^{3}H] (12.5% single-stranded end[a])	10 ml	8.0; phosphate (0.05 M)	40.0
[^{3}H] HeLa (2.2% single-stranded end[a])	5 ml	8.0; phosphate (0.05 M)	4.3
(3.8% single-stranded end[a])	10 ml	8.0; phosphate (0.05 M)	17.0
(8.3% single-stranded end[a])	5 ml	8.0; phosphate (0.05 M)	24.0
(denatured)	30 ml	8.0; phosphate (0.05 M)	9.0

[a]On each side.

Source: Data from Refs. 1 and 75.

extensive bonding occurs when a double-stranded species is denatured to the single-stranded form. Introduction of a single-stranded end into a double-stranded molecule permits covalent attachment to Sepharose. Similarly, the covalent binding of single-stranded RNA can be accomplished effectively. tRNA containing double-stranded regions attaches very poorly by this technique. However, the predominantly double-stranded RNA obtained from reovirus was bound with high efficiency.

Viral DNA experiments indicate, however, that double-stranded DNA can react with the activated gel without single-stranded tails or large single-stranded regions [1,77]. From the temperature dependence of the coupling reaction, it has been concluded [76] that the largest inhibition of coupling of native DNA may result from the inability of the native DNA to penetrate the gel matrix rather than an inherent unreactivity. The Sepharose 2B with a larger pore size showed a lower coupling efficiency than the 4B gel at a reduced temperature. CNBr-activated Sepharose has also been used for the attachment of poly(I·C) [78], poly(U) [79], poly(C), poly(I), poly(sC) [1], and poly d(A·T) [75]. Radioactive poly(I·C) coupled to CNBr-Sepharose retained its biological activity when incubated with chick fibroblasts [81].

It has been suggested that if the condensation reaction with the activated polysaccharide is carried out at pH 6, polynucleotides may be immobilized by single-point attachment at the terminal phosphate groups of the polynucleotide strands [78]. However, in this study, control experiments involving polynucleotides not possessing terminal phosphate groups were not reported, and a more recent investigation of the reaction with CNBr-agarose indicates that multipoint attachment of RNA occurs also at pH 6 [76]. Sites of attachment of poly(U) to agarose gel activated by CNBr were studied using uridine diphosphate and adenine. Independent equilibrium dialysis of A-A for different states of poly(U), in solution and immobilized in gel, were carried out. These results show that the immobilized poly(U) is attached to agarose only at the 5'-terminal phosphate groups and behaves as it does in solution [80]. The immobilization at single-point attachment is probable for the case of poly(U). For polynucleotides having an amino group in the base, the binding reaction occurs at multipoints by the amino groups. The amount of bound nucleotide having an amino group depends on the pH. The amount of bound adenine shows a maximum at pH 9-10. The condition where a polynucleotide having amino groups is bound at a single-point attachment (5' end) may be found by the investigation of the pH

Table 5 Results of RNA Coupling on Different Types of CNBr-Activated and Nonactivated Sepharoses

	Labelled RNA (0.75 mg)			
	[^{3}H]		[^{125}I]	
Matrix (0.25 ml)	cmp × 10^3/ml of gel	Percent coupled	cmp × 10^3/ml of gel	Percent coupled
Sepharose 4B	70	28	575	30
Sepharose 2B	113	45	735	49
ECD-Sepharose 2B	163	65	1100	64
ECD-Sepharose 2B[a]	3.5	1.5	20	1.8

[a]Nonactivated.
Source: Data from Ref. 82.

dependence of the binding reaction of both the 5'-end phosphate group and the amino groups of bases with activated intermediates.

A study has been made of the various parameters that affect the binding mechanism between nucleic acids and CNBr epichlorohydrin-desulfated agarose (ECD-agarose); a hypothetical binding mechanism was proposed [82]. The advantage of ECD-agarose is its ability to resist high temperatures and high concentrations of solvents and chaotropic ions, to remain active over a wide pH range, and to show a high capacity to react with the nucleic acid as ligand. Four different types of matrices were examined for their ability to bind RNA (Table 5). The most efficient binding was obtained with CNBr ECD-Sepharose 2B.

The immobilization of poly(U) on Sepharose via dextran or glycogen spacer has been performed using CNBr activation [83]. The properties of this adsorbent, in which the ligand is separated from the solid matrix by a polysaccharide spacer molecule, were discussed.

B. Azo Coupling

Aryl amine can be coupled to polysaccharides, diazotized, and linked to nucleic acids (Scheme 10). A method for immobilization

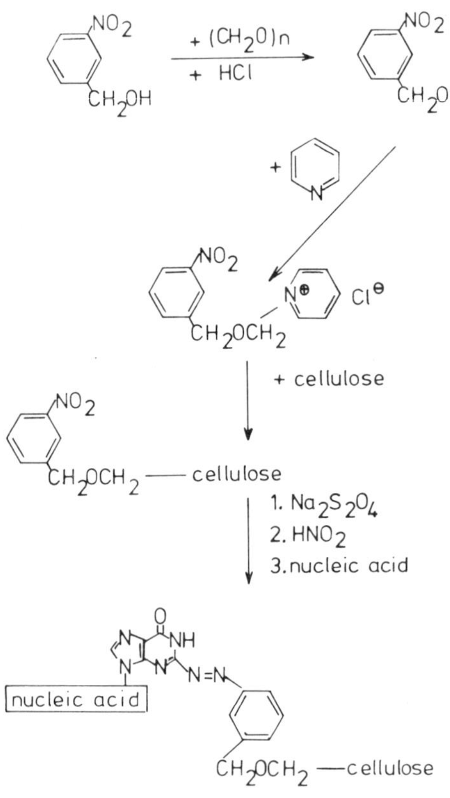

Scheme 10

of DNA and RNA to finely divided cellulose through a diazotized aryl amine, which coupled primarily with guanine and uracil residues of single strands, has been described [84]. Thymidine residues react much more slowly, and no appreciable reaction was observed with polyribocytidine or polyriboadenosine. The high efficiency of coupling and high capacity of the cellulose for nucleic acid make possible a product with as much as 67 μg of nucleic acid per milligram of cellulose. The product is especially suitable for hybridization experiments where very low backgrounds are important, and it is stable in 99% formamide at 80°C, so that hybridized nucleic acid can be recovered easily. Full-length linear SV40 DNA, produced by cleavage of SV40

DNA with S1 nuclease, can be coupled to diazocellulose with an efficiency of 80-90%.

A limited coupling reaction between 4-diazobenzoic acid and sheared single-stranded DNA was employed to prepare a ligand capable of bonding covalently with aminopentane Sepharose [85]. Substantial amounts of DNA (~25 μg/cm^2) can be immobilized to diazobenzyloxymethyl (DBM) paper. Because the DNA is bound to the paper covalently, it is retained through all elution steps. The DNA paper can be reused [86,87]. Since only single-stranded DNA reacts, the coupling procedure is carried out in the presence of a denaturant or after heating the DNA, under conditions where reannealing is slow. Reaction of poly(I) with DBM paper has been accomplished probably through coupling at the 2-position [88]. DBM paper can be used for the transfer of RNAs or DNAs from agarose gels or polyacrylamide-agarose composite gels [89,464].

The synthesis of a family of arylamine-substituted papers that can be diazotized and coupled with nucleic acids has been described [90] (Scheme 11). In the first step of the synthesis an

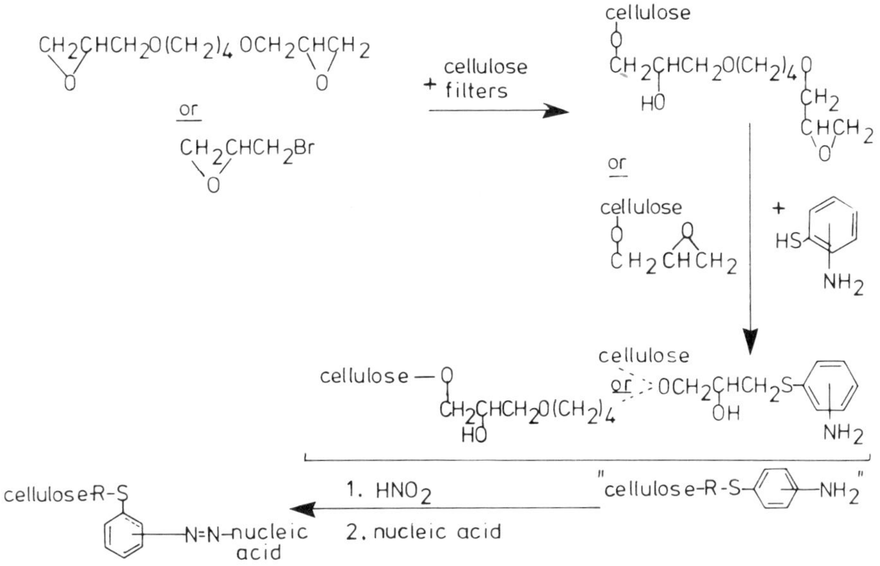

Scheme 11

oxirane-substituted cellulose is created by the reaction of filter paper with a symmetrical bisoxirane or epibromohydrin. In the second step, an unprotected aromatic primary amine is attached to the substrate, taking advantage of the highly nucleophilic character of the thiophenoxide anion. The kinetically favored product is an aromatic amine thioether. Ortho, meta, and para isomers of the aminophenylthioether-coupled cellulose papers were diazotized and reacted with nucleic acids eluted from agarose gels by blotting. The ortho isomer had the highest capacity for DNA in these experiments. The highly labile meta isomer may have been insufficiently stable for good transfer under the conditions employed, while the para form may have been insufficiently reactive. Transfer of labeled DNA fragments to o-DPT paper was found to be relatively insensitive to either pH or ionic strength. The transfer of RNA, however, was favored in low salt. As little as 50 pg of a single species of RNA was detected after electrophoresis in a denaturing agarose gel, transfer by blotting, and hybridization [457].

A defined, covalently immobilized DNA preparation for use in template chromatography has been studied [91]. Alkylation of DNA with 4-bis(2-chloroethyl)amino-L-phenylalanine at a molar ratio of about 1:3 yields over 95% binding of the alkylating agent to the DNA. In a second step the alkylated DNA is coupled to the diazotized Enzacryl AH gel in a yield of 70-80%, that is, at least 5 mg of DNA per 100 g of support. Several controls were performed to show whether the immobilization to the support is via the phenylalanine group and whether the attached DNA retains its double-stranded structure. The results demonstrated that the DNA is irreversibly immobilized to the support by both strands at a limited number of points. The advantages of coupling a ligand to the DNA prior to immobilizing to the support are (a) that the number of linkages can be controlled, and (b) that the ligand will act as a spacer molecule.

C. Condensation by Carbodiimide or Carbonyldiimidazole Activation

Some of the early attempts to immobilize nucleic acids and polynucleotides exploited an activated form of acetylated phosphocellulose. Activation was achieved by treating the derivatized cellulose with dicyclohexylcarbodiimide in a nonaqueous solvent, and it is presumed that the subsequent binding of the polynucleotide resulted from the formation of phosphodiester linkages between the support and the various hydroxyl groups on the polynucleotide chain. Thus it was possible to bind the phosphocellulose to the glucosyl hydroxyl groups of phage T4 DNA [92] and to the 2'-hydroxyl groups of various polynucleotides and RNA molecules [93].

A method has been described which involves specific activation of the terminal phosphate groups of polynucleotides in aqueous solution at pH 6 in the presence of cellulose paper. The activation is brought about by N-cyclohexyl-N'-β-(4-methylmorpholinium)ethyl carbodiimide p-toluenesulfonate [39]. The products of the reaction are considered to consist of polynucleotides connected to the cellulose by ester linkages between the terminal phosphate groups of the polynucleotides and the sterically favored hydroxyl groups of the cellulose. This approach has been adapted to insolubilize sonicated native rat liver DNA on Sephadex G-200 [94]. Sonication of DNA gives rise to double-stranded breaks in the DNA, without loss of its native structure. About 30% of the sonicated DNA bound to Sephadex G-200 under the standard incubation conditions. Increasing the amount of carbodiimide, Sephadex G-200, or DNA in the reaction mixture did not significantly change the percentage of DNA that bound to the Sephadex G-200. Only a small percentage of the sonicated DNA bound to Sephadex G-25 had a total surface area similar to that of Sephadex G-200, under the same incubation conditions. This suggests that the terminal phosphate groups of the DNA are reacting with hydroxyl groups within the matrix of the Sephadex.

A coupling method has been described in which carboxyl groups are introduced into single-stranded DNA by treatment with 4-diazobenzoic acid (Scheme 12). The carboxyl groups introduced in this way were linked to aminopentane-Sepharose using water-soluble carbodiimide [85].

Fragments of denaturated DNA having a length of about 500 nucleotides were immobilized using silica carriers modified by organic polymers. It was shown that the organic coating allows sorbents to be obtained which practically exclude nonspecific sorption with respect to DNA. The immobilization was carried out on hydroxyl-containing sorbents and on their phosphorylated derivatives by means of water-soluble carbodiimide. The resulting preparations contained up to 60 A_{260} units of DNA per gram of carrier [95].

The coupling of double-stranded DNA to cellulose powder has been described [96]. Although double-stranded DNA fragments can readily be linked to the neutral cellulose powder by the carbodiimide method, enzymatically synthesized RNAs (e.g., SV40

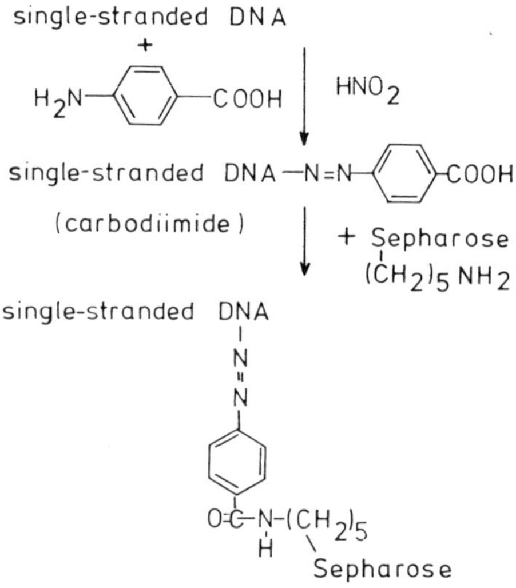

Scheme 12

cRNA), which have fewer nucleophilic triphosphate termini, are more difficult to couple directly to cellulose, especially in the powder form. For natural RNAs, 3'-terminal phosphate groups, generated by partial hydrolytic cleavage during preparation, are rapidly cyclized by carbodiimide and are not linked to cellulose.

RNA has been immobilized by a nonaqueous solid-phase reaction with 1,1'-carbonyldiimidazole and RNA "dry coated" on cellulose or preferably on previously activated phosphocellulose filters. This method does not depend on the terminal groups of RNA chains. Therefore, it has an advantage over the carbodiimide procedure. Immobilization of RNA does not seem to alter its chemical character or cause it to acquire affinity for unspecific RNA or DNA. The immobilized RNA method is well suited for either preparative or analytical purposes (Scheme 13; see also Table 6). This method has been modified to couple RNA covalently to phosphocellulose powder using a two-step carbonyldiimidazole procedure [96]. Immobilization of RNA on phosphocellulose is stable to 15% NH_3 solution, 20 min, 4°C. This suggests that the RNA may be attached at multiple points through phosphoester bonds. Cleavage of the RNA chain at a few sites would therefore not solubilize the linked RNA.

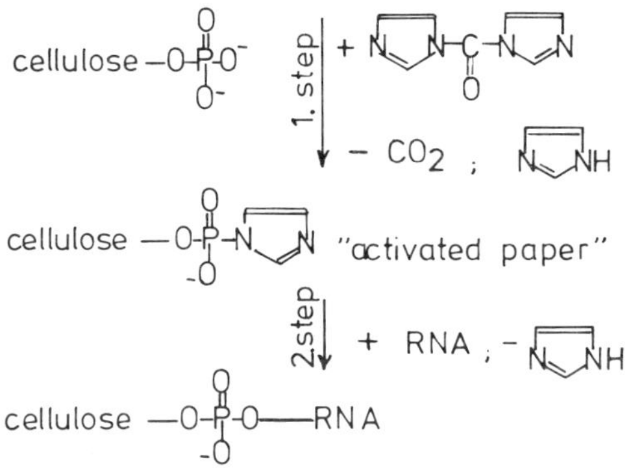

Scheme 13

Table 6 Requirements for Attachment of RNA to Cellulose with Carbonyldiimidazole as Condensing Agent[a]

Ligand	Solvent	Counterion	Filter	Input
Poly(A)	Pyridine	$(Bu)_3NH^+$	Cellulose	4 μg
	Pyridine	$(Bu)_3NH^+$	P-cellulose	4 μg
Pulse-labeled (*E. coli*) RNA	Pyridine	$(Bu)_3NH^+$	Cellulose	1,533 cpm
	Pyridine	K^+	Cellulose	644 cpm
	Pyridine	$(Bu)_3NH^+$	P-cellulose	1,533 cpm
Fibroblast 28S RNA	Pyridine	$(Bu)_3NH^+$	Cellulose	1,200 cpm
	Pyridine	$(Bu)_3NH^+$	P-cellulose	700 cpm
4S Mouse[b] RNA	Pyridine	K^+	Cellulose	32,782 cpm
	Pyridine	K^+	P-cellulose	30,735 cpm

[a]Condensing agent was used at 100 mg/ml (w/v), and the reaction was performed for 8 hr at 37°C.
[b]Cultured L-1210 mouse cells.
Source: Data from Ref. 97.

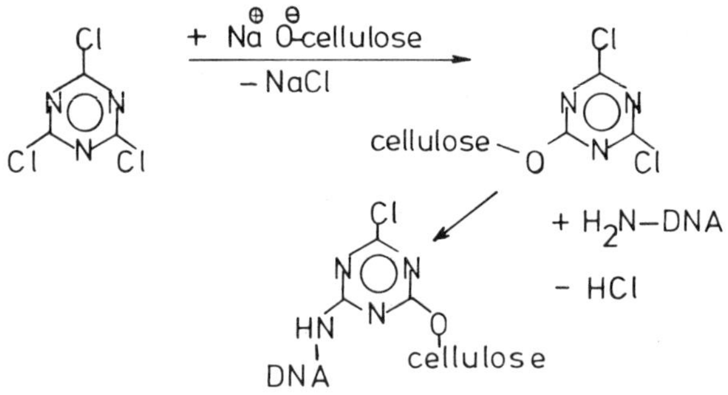

Scheme 14

D. Coupling with Cyanuric Chloride-Activated Cellulose

It has been found that cellulose activated with cyanuric chloride provides a convenient material for the preparation of DNA-cellulose, which has the following favorable characteristics (Scheme 14). It can be prepared by an easy and reproducible method; double-stranded DNA is bound in satisfactory amounts, and the material is very stable, even in urea and in alkaline conditions [98]. This strategy can also be used for the immobilization of nucleic acids to cyanuric chloride-activated paper [99]. The pH value of the binding buffer has a substantial influence on the binding reaction and on the stability of binding. Electrophoretically separated DNA and RNA are transferred to cyanuric chloride-activated paper using an electrotransfer method.

E. Condensation with CM-Cellulose

The coupling of calf thymus DNA to the cellulose derivative is effected by an esterification to the carboxyl functions of the CM-cellulose with terminal hydroxyl groups of the DNA [100]. The standard conditions in the case of coupling commercial calf thymus DNA were 20 mg of DNA per gram of CM-cellulose, pH 3.5, and 40°C, leading to a column material containing 15 mg of DNA per gram of CM-cellulose. The advantages of the material are a rapid and simple preparation procedure, a high yield of bound DNA, and the possibility of linking native as well as denatured DNA to the cellulose. The adsorbents show stability to elevated temperatures and to high formamide concentration but only limited stability to alkaline conditions. Columns can be used repeatedly and allow

Table 7 Results of Coupling by the Mixed Carbonic Anhydride Method

Ligand	Time (hr)	pH; buffer	Amount; % coupled
Adenosine	24	8.5; borate	63 μmol 50.0
Poly(U)	24.0	8.0; borate	30 O.D.[a] 12.5
tRNA	26	8.0; borate	6.75 mg 13.5
DNA calf thymus	22.5	8.0; 0.08 M NaCl	69.4 O.D.[a] 42.0

[a]Optical density.
Source: Ref. 1.

high flow rates. Other products have been coupled to CM-cellulose by the mixed carbonic anhydride method (Tables 1 and 7).

F. Bisoxirane Activation

Activation of Sepharose or cellulose with bifunctional epoxids provides support for the immobilization of denatured DNA and tRNA [101] (Scheme 15). The resulting materials contain the ligand

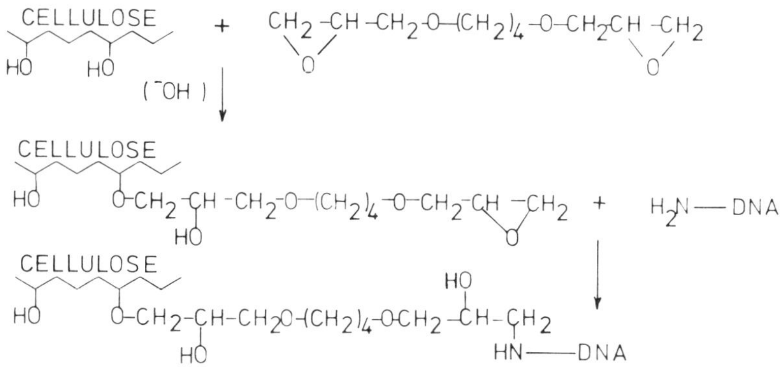

Scheme 15

at the end of a 12-membered, slightly hydrophobic spacer molecule. The efficiency of the approach depends primarily on the pH, temperature, and activity of the matrix. Adsorbents containing ligands immobilized onto cellulose are particularly stable to formamide, temperature, and alkali. However, the stability of Sepharose under such conditions is less evident.

A cloned double-stranded cDNA was immobilized to cellulose using the bifunctional oxirane 1,4-butanedioldiglycidyl ether to activate cellulose [102]. The optimal conditions for the latter reaction included use of a dehydration technique whereby DNA and activated cellulose were allowed to react on glass slides in 0.1 N NaOH. The initial volume of the reaction was important: $\leqslant 250$ μl per 50 mg of cellulose was necessary for maximum efficiency. At DNA concentrations of $\leqslant 4$ μg/mg cellulose, the efficiency of binding was 90%.

G. Efficiency of Different Coupling Procedures

Methods commonly used for covalent immobilization of single-stranded DNA have been applied to several solid supports (Sephadex G-25 and Cellex 410) as well as to a number of macroporous materials (Sepharose Cl-6B and Cl-2B; Sephacryl S-500 and S-1000). Coupling efficiencies and stability of covalently bound DNA were compared for both classes of materials [447]. The yields of the immobilization reaction for sonicated DNA were only 10-40% for G-25 and Cellex 410, in contrast to 60-80% for Cl-6B and S-500. Under optimal conditions, up to 0.5 mg of DNA can be coupled initially per gram of wet macroporous material. The immobilized DNAs are lost from the supports in a biphasic manner, with about 10-20% loss per day during the first 2-3 days at 45°C, followed by only about 1% loss per day at the same temperature thereafter.

The accessibility of immobilized DNA has been shown to depend more crucially on the method of immobilization than on the type of support used for fixation [448]. When sonicated denatured DNA is coupled via diazotization or via cyanogen bromide reaction to solid Sephadex G-25 and Cellex 410 or to macroporous Sephacryl S-500 and Sepharose Cl-6B, its accessibility varies from 100 to 24%. Generally, the loss of accessibility is linked to a depression of the melting temperature of DNA helices formed on the support. This correlation shows a characteristic course for a particular coupling method. DNA coupled under denaturing conditions may become totally inaccessible when only 3% of its bases are involved in the covalent linkage.

III. SPECIAL METHODS FOR THE IMMOBILIZATION OF RNA AND POLYRIBONUCLEOTIDES

The efficiency of immobilization of high-molecular-weight yeast RNA on aminated silica carriers has been studied [449]. More than 90% of the RNA added was bound to the carrier in undegraded form. The procedure allows us to link as much as 2.0 mg of RNA per gram of carrier. During nucleic acid hybridization (65°C, 2 × SSC) in the presence of trace amounts of pronase, chipping off of RNA was observed only during the first 6 hr of incubation. More than half of the immobilized poly(A) RNA remained tightly bound to the carrier. The immobilized messenger and ribosomal yeast RNAs were hybridized with total yeast [^{125}I] DNA. Several methods for the immobilization of polyribonucleotides take advantage of the vicinal diol groups present or potentially present in polyribonucleotides.

A. Immobilization of Oxidized Polyribonucleotides

The free terminal 3'-hydroxyl groups of polyribonucleotides can be oxidized in aqueous solution by periodate as reported in Sec. I.B of Chap. 2 for the oxidation of mononucleotides. The resulting aldehydropolyribonucleotides form Schiff bases with the primary amino groups of AE-cellulose (Scheme 16). The unstable Schiff bases can be converted to stable structures by reduction using $NaBH_4$ [8,103-105]. The binding in these products is thought to arise from the formation of a substituted morpholine ring structure that includes the nitrogen of the AE-cellulose and the five atoms that originally constituted the ribose ring of the terminal nucleoside. The assignment of this linkage is based on the structure of the product obtained from the borohydride reduction of the complex formed between periodate-oxidized AMP and methylamide [106].

The aldehyde-polyribonucleotides may also be added to a hydrazine or a hydrazine-containing support to form a stable hydrazone bond (Scheme 16). For example, a mixture of agar and polyacrylic acid hydrazide has been used in this way to immobilize tRNA [107]. Periodate-oxidized mRNA has been coupled to a linear polymer of acrylic hydrazide entrapped in agar [108]. The reduction with $NaBH_4$ has also been used to stabilize the linkage in a similar complex obtained from periodate-oxidized tRNA and hydrazinyl-Sepharose [109].

The attachment of RNA by its 3' terminus to agarose has been accomplished [110] using alkaline activation of agarose with CNBr,

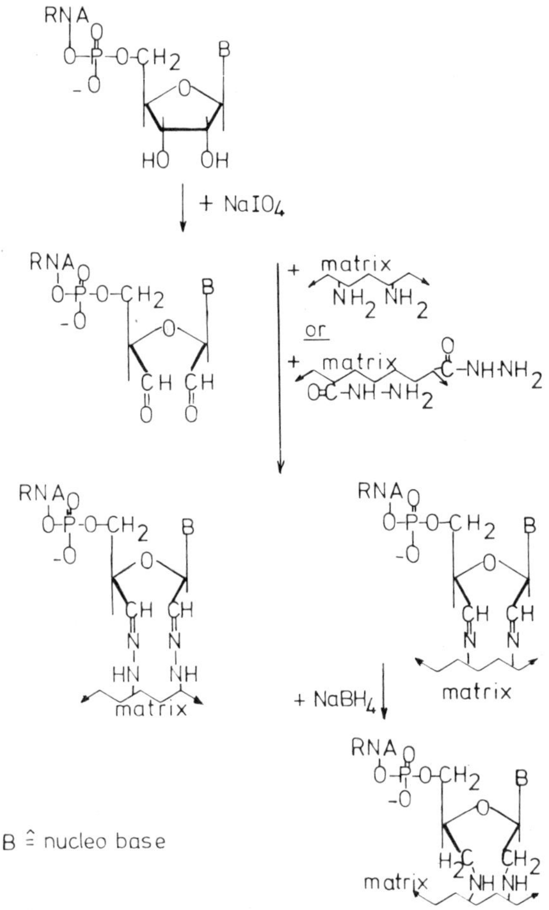

Scheme 16

followed by coupling of aminocarpoic acid methyl ester of the agarose. The caproyl ester agarose was then converted to the corresponding hydrazide. After blocking resin carboxyl groups with glycinamide and water-soluble carbodiimide, the treated agarose is coupled with RNA that had previously been oxidized with sodium periodate. Although rather complicated, this procedure may have specialized use.

On the basis of inorganic materials such as silochrome, porous glass, and chromosorb, a number of supports have been synthesized which have aminopropyl and hydrazide groups. As a result of attachment of periodate-oxidized *Escherichia coli* tRNA to these materials, preparations were obtained having as many as 100-900 A_{260} units per gram of sorbent. To avoid nonspecific sorption that might arise due to excess aminopropyl groups, the preparations of immobilized nucleic acids were treated with acetic anhydride. The supports with immobilized tRNA thus obtained were shown to be almost completely inert with respect to heterological DNA [111].

tRNA and 5S, 16S, and 23S rRNA were readily immobilized on agarose-dihydrazide [112,113], the amount coupled decreasing with increasing molecular size. Epichlorhydrine-activated Sepharose gels were reacted with adipic acid dihydrazide. Oxidized RNA or nucleotides are coupled to hydrazide residues linked by a spacer to the derivatized Sepharose [114].

B. Reversible Complexation of Ribonucleotides with Immobilized Dihydroxyboryl Residues

Terminal cis-diol groups in ribopolynucleotides are capable of forming specific cyclic complexes with supports containing covalently bound dihydroxyboryl groups (Scheme 17). N-[N'-(m-dihydroxyborylphenyl)succinamyl]AE-cellulose prepared from the condensation of N-(m-dihydroxyborylphenyl)succinamic acid with AE-cellulose forms cyclic boronate structures at pH 8-9 with molecules containing the ribonucleoside diol group [115-118]. At pH

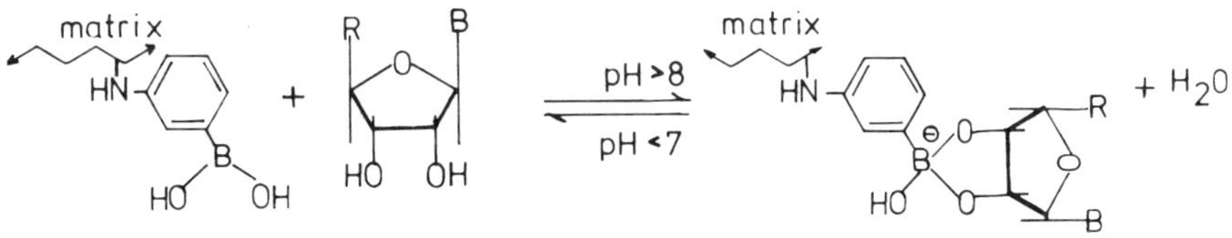

B $\hat{=}$ nucleo base

R $\hat{=}$ CH_2OH or oligonucleotide or ribonucleic acid

Scheme 17

6, the complex breaks down, allowing the recovery of the immobilized ribonucleotides from the support, which is then available for the binding of another ribonucleotides. Similar complex formation can be observed on dihydroxylboryl-substituted methacrylic acid polymer (borate gel) [119]. It should be pointed out that from its chemical constitution, the borate gel appears to be more resistant than the borate-substituted cellulose to degradation by heat or by microorganisms. Moreover, these gels are suitable for chromatography under pressure, which is advantageous for preparative separations. The uses of boronic acids immobilized to insoluble supports include separating ribonucleotides from deoxyribonucleotides [119,120], isolating 3'-terminal polynucleotide fragments from large RNAs [121], and separating 3'-modified polynucleotides from unmodified ones [122]. Immobilized boronic acid can also be used to separate aminoacyl-tRNA from unacylated species [116,119,123-125].

Using dihydroxyboryl cellulose, the 3'-terminal oligoribonucleotides of the in vitro T3 RNA polymerase transcript has been isolated [126]. The advantages and disadvantages of several boronate-containing supports have been investigated [127], with the following results. Boronate cellulose binds macromolecular RNA but often is not useful for trace amounts of material because of nonspecific adsorption. It was found that boronate polyacrylamide and boronate agarose both display very low levels of nonspecific adsorption. Boronate polyacrylamide specifically binds cis-diol-containing mononucleotides or short oligonucleotides, but does not retain polyribonucleotides sufficiently tenaciously for preparative use. Boronate agarose proved to be suitable for the fractionization of trace amounts of macromolecular RNA, but not for mononucleotides.

5

Chromatography of Oligonucleotides

Template chromatography on uracil-coupled cellulose has been used for the separation of nucleosides and mono- and oligonucleotides [140]. Adenine derivatives exhibited a high affinity to uracil-cellulose, and sequential isomers of oligonucleotides containing adenine residues were resolved. This procedure was extended to the isolation of mRNA containing poly(A) sequences. Cellulose-containing oligonucleotides have been investigated with regard to their ability to bind selectively oligonucleotides containing complementary sequences.

Alkali-stable dinucleotides are found in a low proportion (about 3%) in RNA [450]. Partial purification on the milligram scale was achieved by means of a combination of charge-transfer chromatography on acriflavin-Sephadex G-25 and chromatography on Sepharose 4B-RNAase.

Column chromatography of A-A with the immobilized poly(U) was performed at several temperatures and concentrations [80] to analyze the elution profile. The theory of cooperative binding of oligonucleotides to polynucleotides was extended to the framework of plate theory. A computer simulation for the elution profiles was performed using thermodynamic parameters obtained by equilibrium dialysis. This simulation duplicated the experimental results. This fact shows that the peculiar leading form of elution profile is due to the cooperative binding.

Thermal chromatography on columns with oligonucleotide cellulose has been used to resolve mixtures of homooligonucleotides differing by one nucleotide residue in length [133-136]. An extensive examination of the affinity of several series of complementary ribo- and deoxyribooligonucleotides for columns of deoxyribooligonucleotide cellulose has confirmed that all the residues in the cellulose-linked oligonucleotide are able to hydrogen bond to their complementary bases. Oligo(U) were found to form hybrid structures with oligo(dA)-cellulose columns having "melting temperatures" (T_m) lower than those observed for oligo(dT) of equal length. Comparison of oligo(dA) and oligo(A) using columns of oligo(dT)-cellulose again showed interactions with oligo(A) to be less stable (Table 8).

Table 8 Elution of Oligo(dA), Oligo(dT), Oligo(U), and Oligo(A) from Columns of Cellulose-p(dT)$_n$ and Cellulose-p(dA)$_n$

Oligonucleotide cellulose (A.U.[a] incorporated)	Oligonucleotide (A.U.[a] loaded)	Tm (°C)	Oligonucleotide (A.U.[a] loaded)	Tm (°C)
Cellulose-p(dT)$_8$ (27)	p(dA)$_6$ (7.2)	3.5		
	p(dA)$_7$ (6.3)	13.5		
	p(dA)$_8$ (6.2)	19.0		
	p(dA)$_9$ (6.2)	25.5		
	p(dA)$_{10}$ (6.0)	28.5		
Cellulose-p(dT)$_9$ (39)	p(dA)$_6$ (5.4)	8.5	rA(pA)$_5$ (2.4)	4.5
	p(dA)$_7$ (5.5)	18.0	rA(pA)$_6$ (2.5)	12.5
	p(dA)$_8$ (5.5)	26.0	rA(pA)$_7$ (2.0)	18.0
	p(dA)$_9$ (5.6)	32.0	rA(pA)$_8$ (2.0)	25.0
	p(dA)$_{10}$ (5.7)	35.0	rA(pA)$_9$ (4.5)	25.5
	p(dA)$_{11}$ (5.6)	37.0	rA(pA)$_{10}$ (2.5)	26.0
			rA(pA)$_{11}$ (2.7)	28.5
			rA(pA)$_{12}$ (2.0)	29.9
Cellulose-p(dT)$_{12}$ (30)	p(dA)$_6$ (4.0)	8.5	rA(pA)$_6$ (2.5)	9.0
	p(dA)$_7$ (4.8)	14.5	rA(pA)$_7$ (3.0)	15.5
	p(dA)$_8$ (4.2)	20.5	rA(pA)$_8$ (4.0)	22.0
	p(dA)$_9$ (3.0)	27.5	rA(pA)$_9$ (3.0)	27.5
	p(dA)$_{10}$ (4.4)	34.0	rA(pA)$_{10}$ (3.0)	30.0
	p(dA)$_{11}$ (4.8)	37.5	rA(pA)$_{11}$ (3.0)	30.5
			rA(pA)$_{12}$ (1.4)	30.0

Table 8 (Continued)

Oligonucleotide cellulose (A.U.[a] incorporated)	Oligonucleotide (A.U.[a] loaded)		Tm (°C)	Oligonucleotide (A.U.[a] loaded)		Tm (°C)
Cellulose-p(dA)$_8$	p(dT)$_6$	(1.4)	3.0			
(22)	p(dT)$_7$	(1.2)	12.5			
	p(dT)$_8$	(1.5)	18.0			
	p(dT)$_9$	(1.5)	20.0			
	p(dT)$_{10}$	(1.2)	23.0			
	p(dT)$_{11}$	(1.8)	25.5			
Cellulose-p(dA)$_9$	p(dT)$_6$	(4.9)	4.5			
(44)	p(dT)$_7$	(5.4)	14.0			
	p(dT)$_8$	(5.7)	21.0			
	p(dT)$_9$	(5.4)	26.0			
	p(dT)$_{10}$	(5.5)	28.5			
	p(dT)$_{11}$	(5.8)	31.0			
	p(dT)$_{12}$	(5.3)	33.5			
Cellulose-p(dA)$_{10}$	p(dT)$_6$	(2.0)	4.0			
(49)	p(dT)$_7$	(2.0)	17.0	(rUp)$_7$	(4.0)	-4.0
	p(dT)$_8$	(2.0)	24.0	(rUp)$_8$	(4.0)	-3.0
	p(dT)$_9$	(2.0)	29.5	(rUp)$_9$	(3.3)	3.5
	p(dT)$_{10}$	(2.0)	33.5	(rUp)$_{10}$	(5.0)	8.5
	p(dT)$_{11}$	(2.0)	34.5	(rUp)$_{11}$	(4.1)	12.0
	p(dT)$_{12}$	(2.0)	36.5	(rUp)$_{12}$	(5.7)	15.5
	p(dT)$_{13}$	(2.0)	38.5	(rUp)$_{13}$	(4.8)	18.5
				(rUp)$_{14}$	(4.1)	20.0
				rU(pU)$_7$	(1.1)	-2.0
				rU(pU)$_8$	(0.9)	6.0
				rU(pU)$_9$	(1.0)	10.0
				rU(pU)$_{10}$	(1.2)	14.0
				rU(pU)$_{11}$	(0.9)	15.0
				rU(pU)$_{12}$	(1.0)	16.0
				rU(pU)$_{13}$	(1.0)	22.0

[a]As defined in Ref. 40.

Source: Reprinted with permission from Ref. 135. Copyright 1973 American Chemical Society.

A study of the interactions of a series of deoxyribooligonucleotides of mixed repeating base sequence with columns of complementary oligonucleotide cellulose has been made (Table 9). The stability of the interactions of oligo(A), containing single nucleotide substitutions, with cellulose-p(dT)$_9$ has been studied by thermal elution (Table 10). In the case of oligo(dA), the replacement of an internal dA by dC/dG or dT causes destabilization. In the case of oligo(A), replacement of an internal A residue by C or U resulted in a similar destabilization. However, replacement of an internal A residue by G resulted in significantly less destabilization. These false monomer units in the complementary oligonucleotide lcwer its melting temperature. In template chromatography oligonucleotides with a low content of false sequences are thus eluted before the completely complementary oligonucleotides during the temperature gradient. The results indicate that caution should be exercised in the extrapolation of data on the specificity of deoxyribopolynucleotide-deoxyribopolynucleotide hybridization to structures of the deoxyribopolynucleotide-ribopolynucleotide type.

The order of elution may be explained by the base-pairing mechanism as follows. Complementary oligonucleotides strive to form base pairs under certain conditions. The degree of hybridization increases with the number of complementary paired bases within the nucleotide chain. An increase in temperature reduces the degree of base pairing, as the secondary valence bonds of the paired bases become unstable. At temperature above the "melting point," base pairing of short complementary oligonucleotides disappears, whereas longer oligomers remain hybridized. According to the temperature gradient to the column during template chromatography, the adsorbed oligonucleotides desorb in the order of the number of their paired monomer units. The melting points of complementary oligonucleotides shift to higher temperatures with an increasing number of paired bases, but at the same time the difference in melting point for each homolog unit decreases. As a result, the longer-chain oligonucleotides cannot be fully resolved. In addition, the melting range widens, apparently because the oligonucleotides of the stationary phase are not uniform in their molecular structure or are not accessible to the same degree. Thus partial hybridization between mobile and immobilized complementary oligonucleotides is favored. Partially hybridized oligonucleotides "melt" at lower temperatures than do fully hybridized ones. Their partial hybridization can lead to simultaneous elution of oligonucleotides of higher chain length with lower homologs which undergo complete hybridization. Even if the stationary

Table 9 Chromatography of Deoxyribonucleotides of Mixed Repeating Base Sequence on Columns of Complementary Deoxyribooligonucleotide Cellulose[a]

Interaction of base pairs	T_m (°C)[b]
5' ——— 3' Cellulose–pdT-T-C-T-T-C · · · · · dG–A-A-G-A-Ap or 3' ——— 5' Cellulose–pdT-T-C-T-T-C · · · · dG-A-A-G-A-Ap	Retarded but not retained at -5°C
Cellulose–pdT-T-C-T-T-C · · · · · · dG-A-A-G-A-A-G-A-Ap	15.5
Cellulose–pdC-T-T-C-T-T · · · · · · dG-A-A-G-A-Ap	14.0
Cellulose–pdC-T-T-C-T-T · · · · · · dG-A-A-G-A-A-G-A-Ap	19.0
Cellulose–pdT-T-C-T-T-C-T-T-C · · · · · · dG-A-A-G-A-Ap	4.5
Cellulose–pdC-T-T-C-T-T-C-T-T · · · · · · dG-A-A-G-A-Ap Cellulose–pdC-T-T-C-T-T-C-T-T · · · · · · dG-A-A-G-A-Ap	16.0
Cellulose–pdT-T-C-T-T-C-T-T-C · · · · · · · · · dG-A-A-G-A-A-G-A-A	33.5
Cellulose–pdC-T-T-C-T-T-C-T-T · · · · · · · · · dG-A-A-G-A-A-G-A-Ap	40.5

[a]The column (5 cm × 0.9 cm) contained the covalently linked oligonucleotide: p(dT-T-C)$_2$, 92 A_{260} units; p(dC-T-T)$_2$, 97 A_{260} units; p(dT-T-C)$_3$, 77 A_{260} units; p(dC-T-T)$_3$, 103 A_{260} units. Complementary oligonucleotides 2.5 to 3.0 A_{260} units were loaded at -5°C in 1 ml of 1 M sodium chloride-0.01 M sodium phosphate (pH 7.0) and eluted in the same buffer using a linear temperature gradient.

[b]As defined in Ref. 40.

Source: Reprinted with permission from Ref. 135. Copyright 1973 American Chemical Society.

Table 10 Effect of Substitution of a Single Nucleoside in a Oligo(dA) of the Type pdA_5-dN-dA_n on the T_m (°C) Values with Cellulose-pdT_9 in 1 M NaCl at pH 7.0

pdA_6	7	pdA_5-dC	a	pdA_5-dG	a	pdA_5-dT	a
pdA_7	18.5	pdA_5-dC-dA	6	pdA_5-dG-dA	4	pdA_5-dT-dA	6.5
pdA_8	26	pdA_5-dC-dA_2	13	pdA_5-dG-dA_2	9.5	pdA_5-dT-dA_2	11.5
pdA_9	31	pdA_5-dC-dA_3	16	pdA_5-dG-dA_3	15.5	pdA_5-dT-dA_3	15.5

Effect of the Position of a Deoxyriboguanylate Residue in a Oligo(dA) on the T_m (°C) Values with Cellulose-pdT_9 in 1 M NaCl at pH 7.0

pdA_6	7					pdA_4-dG-dA	a
pdA_7	18.5			pdA_6-dG	11	pdA_4-dG-dA_2	a
pdA_8	26	pdA_7-dG	22	pdA_6-dG-dA	18	pdA_4-dG-dA_3	5.5
pdA_9	31			pdA_6-dG-dA_2	23		

Effect of Substitution of a Deoxyribonucleoside at the 3' Terminus of Deoxyribooligoadenylates on the T_m (°C) Values with Cellulose-pdT_9 in 1 M NaCl at pH 7.0

pdA_6	7						
pdA_7	18.5	pdA_6-dC	6	pdA_6-dG	11	pdA_6-dT	8
pdA_8	26	pdA_7-dC	16	pdA_7-dG	22	pdA_7-dT	16.5

Effect of Substitution of a Single Nucleoside in a Oligo(A) of the Type A_4-N-A_n on the T_m (°C) Values with Cellulose-pdT_9 in 1 M NaCl at pH 7.0

A_6	3	A_4-C-A	a	A_4-G-A	a	A_4-U-A	a
A_7	11	A_4-C-A_2	a	A_4-G-A_2	7	A_4-U-A_2	a
A_8	18.5	A_4-C-A_3	a	A_4-G-A_3	15.5	A_4-U-A_3	1
A_9	24	A_4-C-A_4	1	A_4-G-A_4	19.5	A_4-U-A_4	7.5
		A_4-C-A_5	6.5				

Effect of a G Residue at the 3 Terminus of a Oligo(A) on the T_m (°C) Values with Cellulose-pdT_9 in 1 M NaCl at pH 7.0

A_6	3	A_5-G	a
A_7	11	A_6-G	7.5
A_8	18.5	A_7-G	16
A_9	24	A_8-G	19.5

[a]Not adsorbed at -4°C.

Source: Ref. 136.

phase contains only uniform oligonucleotide molecules, partial hybridization caused by steric hindrance cannot be excluded. According to our experience, the greater effort necessary for the preparation of uniform oligonucleotides covalently bound to the support matrix is not compensated by the improved efficiency of separation.

Oligonucleotide-DEAE-cellulose [47-49] is particularly suitable for use in preparative template chromatography, since in contrast to oligonucleotide-cellulose, its preparation is readily carried out on a large scale. Synthetic homopolymers of thymidylic acid and deoxyadenylic acid with up to 13 monomer units have been chromatographed with a temperature gradient on both oligo(dT)- and oligo(dA)-DEAE-cellulose [48,49]. It is found that both linear oligonucleotides and pyrophosphate derivatives are adsorbed on the immobilized complementary oligonucleotides, whereas no base pairing is observed for noncomplementary or for cyclic oligonucleotides. The strength of base pairing increases with increasing chain length of the oligonucleotide and decreases with increasing salt concentration. Mixtures of low-molecular-weight linear homopolymers are separated into a fraction of uniform molecular weight by the base-pairing mechanism. Pyrophosphate derivatives compete with linear oligonucleotides in base pairing and complicate the separation. Mixtures of oligonucleotides of high molecular weight are eluted as partially resolved fractions. The elution of the oligonucleotides according to molecular weight is performed with a temperature gradient from 0 to 30°C.

DNA can be partially hydrolyzed in a preparative scale to oligomers of adenylic, thymidylic, or guanylic acid by chemical methods. Isolation of homologs with six or more monomer units, which is either very difficult or impossible using the usual chromatographic methods, can be achieved with template chromatography.

For example, the alkaline hydrolysis of oxidized DNA from herring sperm yields a complex mixture of oligo(dA). After the removal of the fragments containing one to three monomer units by column chromatography, approximately 10% of the partial hydrolysate remains. This remaining fraction, which contains the fragments of higher molecular weight, is separated into two fractions by the base-pairing mechanism on a PVAL$(pT)_n$-DEAE-cellulose column with a two-step temperature gradient (Fig. 4). The mixture of oligonucleotides is placed on a PVAL-p$(dT)_n$-DEAE cellulose column at 30°C. After cooling the column to -4°C, 1 M NaCl and 0.01 M Na_2HPO_4 (pH 6.5) at the same temperature are used for elution until absorption drops below 0.04 A_{260} unit. Afterward, the column is warmed to 30°C and the same buffer, now at 30°C, is used for elution. In peak I of Fig. 4 all compounds (~95%) are eluted which do not form base pairs with the immobilized oligo(dT). The compounds (~5%) adsorbed from the mobile phase at -4°C, however, are desorbed only after raising the temperature to 30°C, and are eluted in peak II of Fig. 4. The products in peak II are desalted, enzymatically dephosphorylated,

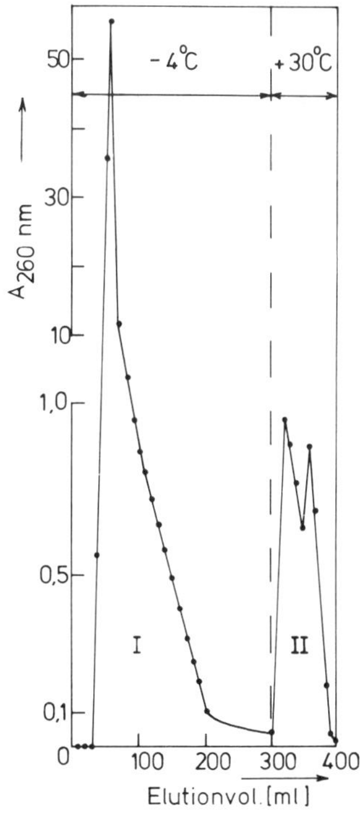

Figure 4 Elution profile of about 1200 A_{260} units of the fraction, which contained high-molecular-weight oligo(dA) isolated from the partial hydrolysate of oxidized herring sperm DNA. The separation is carried out on a PVAL-p(dT)$_n$-DEAE-cellulose column (20 × 2 cm) in a two-step temperature gradient. Flow rate: 40 ml/hr. The products of peak II are combined, desalted, lyophilized, enzymatically dephosphorylated, and rechromatographed (see Fig. 5). (From Ref. 137.)

and fractionated according to the number of monomer units on QAE-Sephadex (Fig. 5). In this manner homologs of deoxyriboadenylic acid with up to eight monomer units are obtained in preparative amounts.

In another approach herring sperm DNA (500 g) is chemically degraded to oligo(dT), which is subsequently fractionated on a QAE-Sephadex column according to the increasing charges of their phosphate groups. The final fraction contains a mixture (1.5 g) of long-chain oligo(dT). Chromatographically pure, single substances (1-4 mg) of the $(dT)_{3-12}$ series are separated from the mixture using template chromatography and rechromatography on QAE-Sephadex [138]. Oligo(dG) [139] which is isolated from the partial hydrolyzate of herring sperm DNA is separated by template chromatography on two differently substituted matrices of PVAL-$(dC)_n$-DEAE-cellulose.

Purine oligonucleotides are adsorbed at 0°C on PVAL-p$(dC)_n$-DEAE-cellulose, and pyrimidine oligonucleotides on oligo(guanylic acid) gel [46] according to the base-pairing mechanism, if their

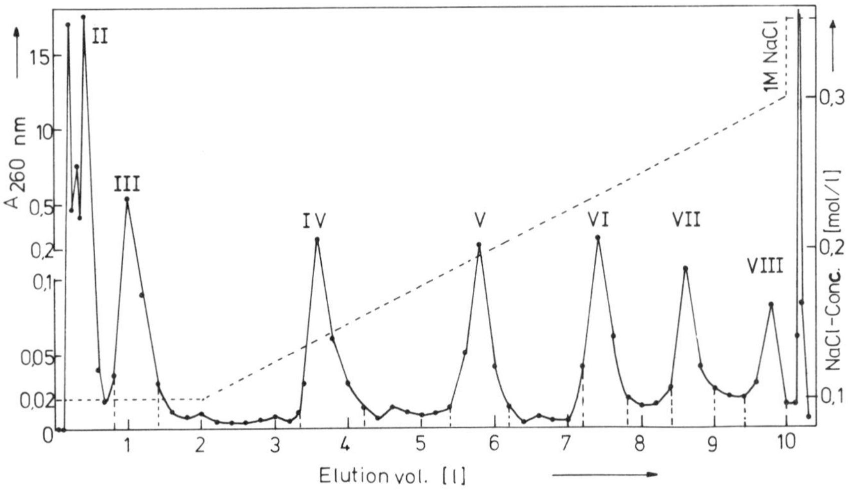

Figure 5 Elution profile of the rechromatography of the products that are eluted in peak II of Fig. 4. About 940 A_{260} units are rechromatographed at 20°C on a QAE-Sephadex A 25 column (47 × 2 cm) using a NaCl gradient. Peaks II to VIII contain the homologs of the $(dA)_{2-8}$ series. (From Ref. 137.)

sequences contain at least three or more homologous, consecutive guanylic or cytidylic moieties [128]. By increasing the temperature, all base-paired oligonucleotides are desorbed. With this template chromatography, mixtures of defined purine or pyrimidine oligonucleotides could be isolated from fractionated partial hydrolyzates of herring sperm DNA. Afterward, these mixtures are rechromatographed on QAE-Sephadex and/or Nucleosil C_{18}. Using this approach, oligonucleotides of up to nine monomer units can be isolated either as single substances or as mixtures with defined composition on a preparative scale, which would be not possible with the already existing separation procedures when partial hydrolyzates of herring sperm DNA are used as starting material. The purity and sequence of the isolated oligonucleotides are determined by the "fingerprint" method.

6

Isolation of DNA and Its Fragments

I. SIMIAN VIRUS 40 DNA

Complementary [^{3}H]-labeled SV40 DNA could be quantitatively isolated in the presence of an excess of unrelated *Micrococcus lysodeikticus* DNA utilizing a phosphocellulose column containing unlabeled SV40 cRNA, connected in series to a denaturation column [96]. Full-length linear SV40 DNA immobilized to diazocellulose is effective in hybridization experiments with SV40 DNA, complementary RNA synthesized in vitro from SV40 DNA with *E. coli* RNA polymerase, and the SV40-specific fraction of total RNA from SV40-infected and transformed cells [84].

II. DNA CONTAINING OLIGO(dG·dC) CLUSTERS

DNA containing oligo(dG·dC) clusters can be isolated by specific binding to poly(C)-Sephadex. This binding is probably mediated by the formation of an oligo(dG·dC)·C triple helix [142]. It has been shown that the viral DNAs PM2, lambda, and SV40 contain at least one, one and two sites for binding to poly(C)-Sephadex, respectively. Inspection of the sequence of the bound fragments of SV40 DNA shows that a $(dG \cdot dC)_{6,7}$ tract is required for the binding of duplex DNA to poly(C)-Sephadex.

A method that allows the rapid preparation of oligo dG-tailed plasmid vectors has been presented [461]. The procedure involves purification of the tailed molecules by hybridization to oligo(dC)-cellulose followed by a stepwise thermal elution. The resulting plasmid is virtually devoid of transformation activity in the absence of oligo(dC)-tailed DNA fragments.

III. DNA COMPLEMENTARY TO GLOBIN RNA

DNA fragments containing sequences complementary to globin mRNA were isolated from pigeon reticulocytes using the binding of the poly(A) sequence of the globin mRNA·DNA hybrids to poly(U)-Sepharose [143]. The length of DNA complementary to a rabbit globin mRNA template was determined and the structure of its 3' end investigated [144]. The results showed that at least one-fifth to one-fourth of the DNA molecules are fully transcribed and contain a poly(dA) sequence at the 3' end.

The cloning of DNA sequences in plasmid recombinants has made it possible to amplify specific sequences to an extent that they can be used for preparative purposes. The use of rabbit globin DNA sequences cloned in the plasmid pCRI and covalently bound to Sepharose 4B for the purification of chain-specific rabbit α- and β-globin cDNAs is described [145]. These purified probes were then used to estimate the length of the α- and β-globin DNA sequences inserted into the recombinant plasmid.

Human globin mRNA was prepared from human reticulocytes and RNA was labeled in vitro by iodination with ^{125}I. Fragmented DNA was hybridized with human globin mRNA and purified by chromatography [146]. The yield of bound material was 0.5% on poly(U)-Sepharose and 0.01-0.04% on SH-Sepharose. SH-Sepharose chromatography is more efficient than poly(U)-Sepharose chromatography for hybrid recovery because it gives lower background.

IV. FRAGMENTS OF DIFFERENT DNAS

Fragments of yeast mtDNA have been fractionated on poly(U)-Sephadex columns [147]. The poly(U) forms a triple helix with (dA·dT) clusters in duplex DNA and fractionates DNA fragments on the basis of the length and number of clusters contained in them. mtDNA fragments obtained with endonucleases were separated by poly(U)-Sephadex in three groups: fragments not retained by the column in 2 M LiCl, fragments partially retained, and fragments (nearly) completely bound in 2 M LiCl and eluted only by 0.1 M LiCl. The separation obtained is adequate for analytical fractionation of fragments and it can be used for the preparative isolation of firmly bound fragments. In mtDNA digests made with endonuclease HapII, which gives about 70 separable fragments under these conditions, only about 10% of the fragments were firmly bound to poly(U)-Sephadex. This shows that the number of (dA·dT) clusters long enough to result in binding is limited in yeast mtDNA and it suggests that large fragments are bound by only one or a few clusters. Corresponding segments of the physical map of the mtDNAs from *Saccharomyces carlsbergensis* and *S. cerevisiae* strains JSI-31 and KL14-4A were bound to the column, showing that the (dA·dT) clusters responsible for binding are conserved in the evolution of mtDNA.

A method for the rapid isolation of mtDNA from the yeast *S. cerevisiae* has been described [462]. Cells are first disrupted by vortexing with glass beads and the mtDNA is then extracted directly from the cell lysate by poly-L-lysine-kieselguhr-exchange chromatography. The method is unique from most other published procedures in that there is no requirement for the isolation of either a crude or purified mitochondrial preparation. mtDNA isolated by this procedure was shown to yield restriction endonuclease fragment patterns identical to those obtained from DNA isolated by procedures reported previously.

Separation of native DNA fragments, with structural genes at the end, was carried out by obtaining fragments of native DNA of a definite size, splitting a small section of 3'-terminal chains from these fragments with *E. coli* exonuclease III, hybridizing poly(A)$^+$ mRNA with these sections, and separating the hybridized DNA fragments on poly(U)-Sepharose [148].

A procedure for enriching DNA for specific sequences based on R looping has been developed [149]. R loops are formed with the DNA using mRNAs containing the sequence of interest and then isolated on poly(U)-Sepharose via the poly(A) tail of the mRNA. Model experiments showed that plasmid DNA containing a cDNA copy of an immunoglobulin kappa chain mRNA could be selectively retrieved using this procedure. Approximately 5-10% of the kappa sequences in mouse embryo DNA could be recovered by R looping, while nonspecific binding of mouse DNA to the poly(A)-Sepharose column was 0.03-0.04%. This represents a 100-to-200-fold enrichment of mouse genomic kappa sequences. The procedure was also used to screen a mouse clone library rapidly for immunoglobulin heavy-chain genes.

The two major rapid DNA sequencing methods in current use are the chemical degradation method and the chain terminator method. The latter has the advantage of being less laborious. However, it requires single-stranded DNA as a template, and methods for strand separation are not readily available for DNA cloned into plasmid vehicles. One approach to this problem is to make a cloning vehicle whose strands can easily be separated. Such a plasmid has been constructed which contains a poly(dA)· poly(dT) duplex segment of length approximately 100 base pairs inserted into the PvuII site of pBR322. The complementary strands of the linearized denatured plasmid DNA can be separated rapidly and efficiently using chromatography with oligo(dA)- and oligo(dT)-cellulose columns in series [150]. More than 90% of the input DNA is recovered as separated strands which can be annealed to form full-length double-stranded molecules.

V. DIAGNOSIS OF POTATO SPINDLE TUBER VIROID DISEASE

A sensitive and reliable method for the detection of potato spindle tuber viroid in potato tubers has been developed [151]. The method is based on hybridization of highly radioactive recombinant DNA to viroid RNA that has been attached to a solid support. The method can be automated and permits the rapid testing of large numbers of tubers.

VI. IDENTIFICATION OF 5-METHYLCYTOSINE IN DNA

A method to identify 5-methylcytosine (m^5Cyt) in DNA [152] immobilized on nitrocellulose paper by using antibody against m^5Cyt raised in rabbits has been described. Immobilized restriction fragments of DNA are incubated first with purified antibody against m^5Cyt and then with goat antirabbit IgG labeled with ^{125}I. Restriction fragments containing m^5Cyt are visualized by autoradiography. A heavily methylated fragment of about 1700 base pairs was identified in nuclear DNA from Chinese hamster cells, the methylation pattern of calf thymus satellite I DNA was examined, and chloroplast DNAs that were extracted from various stages of the *Chlamydomonas* life cycle were compared. Little if any methylation was detected in chloroplast DNA from vegetative cells or from male gametes, whereas homologous DNAs from female gametes and from zygotes were heavily methylated. The presence of m^5Cyt was detected with as little as 50 ng of ϕX174 DNA containing 0.02 pmol of m^5Cyt and with 100 ng of satellite DNA containing about 0.5 pmol of m^5Cyt. Thus the method makes possible the identification of individual methylated sites in purified DNAs in the size range of single genes.

VII. PURIFICATION OF THE OVALBUMIN GENE

Cellulose-bound DNA complementary to ovalbumin mRNA was used in a continuous hybridization system to isolate single-stranded DNA molecules containing the ovalbumin gene [453]. Fragmented DNA segments containing the ovalbumin gene were enriched 300-to-350-fold in one cycle of purification. Two cycles of purification resulted in a DNA fraction which was enriched 2300-fold in the ovalbumin sequence.

DNA complementary to purified ovalbumin mRNA was synthesized in vitro. The cDNA was covalently coupled to phosphocellulose previously activated by N,N'-carbonyldiimidazole. The cDNA-phosphocellulose [454] was employed in a recycling system to enrich the complementary DNA sequence from total chick DNA sheared to an average chain length of 5000 base pairs. The chick DNA sequences complementary to the cDNA (the anticoding strand of the ovalbumin gene) were bound to the cDNA-phosphocellulose column.

A technique has been described for transferring electrophoretically separated bands of double-stranded DNA from agarose gels to DBM paper [567]. Controlled cleavage of the DNA in situ by sequential treatment with dilute acid, which causes partial depurination, and dilute alkali, which causes cleavage and separation of the strands, allows the DNA to leave the gel rapidly and completely, with an efficiency independent of its size. Covalent attachment of DNA to paper prevents losses during subsequent hybridization and washing steps and allows a single paper to be reused many times. Ten percent dextran sulfate, originally found to accelerate DNA hybridization in solution by about 10-fold, accelerates the rate of hybridization of randomly cleaved double-stranded DNA probes to immobilized nucleic acids by as much as 100-fold, without increasing the background significantly.

7

Isolation of RNA

Affinity techniques have become important methods for the isolation of specific RNA sequences [460]. Chromatography on columns of oligo(dT)-cellulose or poly(U)-agarose separates most mRNAs containing 3'-terminal poly(A) segments from total cellular RNA (Fig. 6). Polysome immunoprecipitation, which employs antibodies directed against native antigens to bind to nascent peptide chains on ribosomes, effectively isolates specific polysomes with their mRNAs (Fig. 7). Enzymatic synthesis of cDNAs on oligo-(dT)-cellulose provides reusable hybridization affinity probes for specific sequences as well as mRNA populations. Chemical coupling of mRNAs or various DNA fragments to supports is an important approach to developing specific solid-phase affinity ligands.

I. VARIOUS POSSIBILITIES OF THE RNA ISOLATION

Immobilized DNA forms specific hybrids with complementary RNA. This provides a general method for the isolation of RNA having sequence homology with DNA from any given source. The method is illustrated by a study of the rate of synthesis of complementary RNA in *Proteus vulgaris* [64].

An improved method for the formation of DNA·RNA hybrids has been described [68]. The procedure involves immobilizing denatured DNA on nitrocellulose membrane filters, hybridizing complementary RNA to the membrane-fixed DNA, and eliminating RNA "noise." Denatured DNA and hybridized RNA remain on the

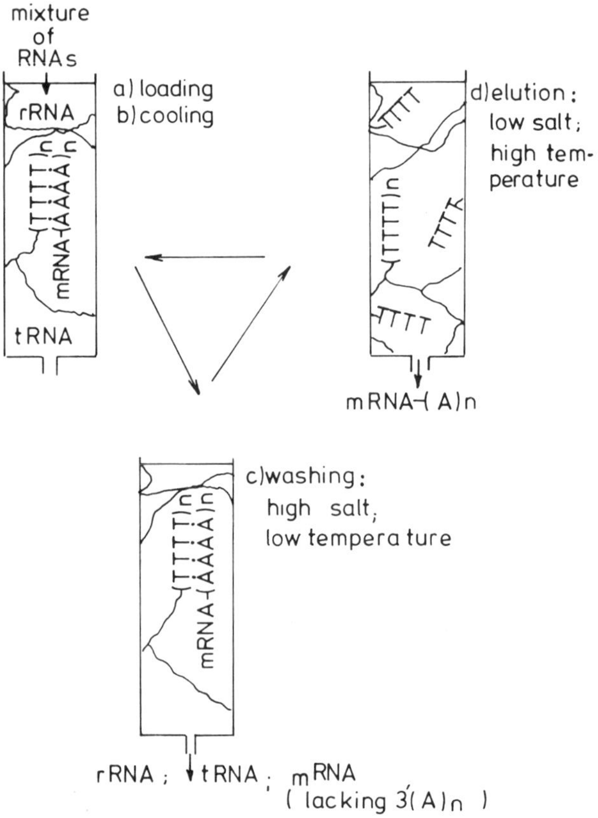

Figure 6 Separation of mRNA from a mixture of RNAs using oligo-(dT)-cellulose column. A poly(U)-Sephadex column can be used instead of the oligo(dT)-cellulose.

filter throughout the procedure. Unpaired RNA is removed by washing, and RNA complexed over short regions is eliminated by RNase treatment. Many samples can be easily handled, permitting kinetic and saturation studies. Large amounts of DNA can be loaded on a filter without interfering with the efficiency of the annealing reaction. Moreover, the "noise" can be depressed to a level (0.003% of the input RNA) permitting the identification of small regions of DNA complementary to a given RNA species. The

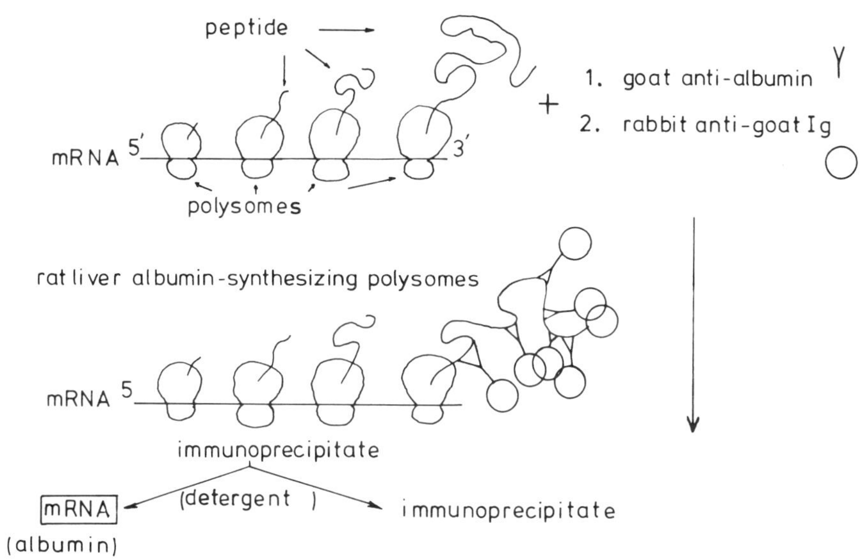

Figure 7 Isolation of specific polysomes with their mRNA.

results are quantitatively more certain than annealing in liquid, since the competing DNA renaturation reaction is suppressed. A guanine derivative of cellulose adsorbed polynucleotides at high ionic strengths. Yeast RNAs were separated into fractions of different base composition [33].

Substantial amounts of DNA (at least 25 $\mu g/cm^2$) can be covalently bound to DBM paper [86]. Complementary RNA will hybridize to the DNA paper almost completely in 24 hr. Using several different conditions of hybridization and washing, the background of RNA bound nonspecifically is between 0.1 and 0.02% and the efficiency of hybridization is very high; 75 to 80% of complementary RNA is bound and retained through the washing procedure.

Biotin-labeled DNA probes, prepared by nick translation in the presence of biotinylated analogs of TTP, are hybridized to DNA or RNA immobilized on nitrocellulose filters [509]. After removal of residual probe, the filters are incubated for 2-5 min with a preformed complex made with avidin-DH and biotinylated polymers of intestinal alkaline phosphatase. The filters are then incubated with a mixture of 5-bromo-4-chloro-3-indolyl phosphate and nitro blue tetrazolium, which results in the deposition of a purple precipitate at the sites of hybridization. This procedure will detect target sequences in the range 1-10 pg after enzyme incubation periods of 1 hr or less. This nonradiographic method of probe detection should be of general utility for genetic studies using Southern, RNA, or dot-blot hybridizations protocols.

Procedures have been developed for the isolation and purification of polynucleotide fragments deriving from the terminals of large RNA [105]. The method depends on the selective periodate oxidation of the 2',3'-diol groups of the terminal fragment after cleavage of the RNA, and the specific binding of this fragment to AE-cellulose. After the unoxidized polynucleotide fragment is released by a β-elimination reaction and is thus isolated in a form which lacks its terminal nucleoside. The fragments obtained from the bacteriophages f2 and Qβ have been characterized by alkaline and enzymatic hydrolyses and shown to have the compositions $(A_2U_2C_4)$-Cp and (C_9U_4)-Cp, respectively.

Double-stranded cDNA was made to partially purified mRNA for the small seminal vesicle secretory proteins IV and V [481]. This cDNA was then inserted into the Pst 1 site of pBR322 by the (G-C) homopolymer tailing technique. Bacterial transformants harboring plasmids with specific inserts were identified by translation of mRNA that was hybridized to plasmid DNA immobilized on nitrocellulose. Separate plasmids were obtained with cDNA inserts for both SVS IV and V. Neither hybridization results nor preliminary restriction analysis gave any indication for homology between them.

An individual mRNA species was isolated directly from lysates of *Physarum polycephalum* by hybridization to immobilized complementary DNA [508]. The solvent for lysis was a 6 M guanidinium salt solution, and the hybridization reaction was carried out at 42°C in 4 M guanidinium salt/33% (v/v) formamide buffer. The mRNA species isolated by this procedure was active in cell-free protein synthesis.

RNA complementary to kinetoplasts of *Crithidia luciliae* was labeled with fluorescein and hybridized with Sephadex beads to which kinetoplast DNA or heterologous DNA had been covalently bound as well as to *C. luciliae* preparations. The fluorescein-labeled RNA was found to hybridize specifically with homologous DNA both on the beads and in the cells [154]. The sensitivity of the hybrid detection could be increased by applying an indirect immunofluorescence reaction using rabbit antiserum raised against the hapten fluoresein.

Low-molecular-weight RNA fractions were isolated from rat calvaria by molecular sieve chromatography. One of these fractions contained tracts of poly(U) and was further purified by chromatography on oligo(dA)-cellulose [155].

Mixtures of high-molecular-weight RNAs have been separated on a preparative scale by high-pressure liquid chromatography (HPLC) gel filtration [451]. The fractionation limit of RNAs was around 42S in nondenaturing solvent systems, but decreased to approximately 16S under denaturing conditions in the presence of 6 M urea. The use of denaturing elution buffers improved resolution considerably. Translational activity of globin mRNA and of mRNAs of vesicular stomatitis virus was retained after gel permeation chromatography.

The vegetative prothalli (1-3 weeks old) of *Anemia* were incubated for 24 hr in [^{14}C]adenine. The RNA was phenol extracted from whole cells and the poly(A) sequences were isolated by nuclease digestion followed by poly(U)-Sepharose chromatography [153].

Three fractions of poly(A) RNA were separated from total rat liver RNA using poly(U)-Sepharose 4B chromatography [156]. The poly(A) RNA fractions were released by thermal elution.

Rat brain poly(A) RNAs were prepared by chromatography on poly(U)-Sepharose and translated in the rabbit reticulocyte lysate system in the presence of [^{35}S]methionine [477]. By using antimyelin proteolipids antibodies, immunoreactive polypeptides of 23.5 and 19 kb were isolated from the translation products: they had electrophoretic mobilities similar to those of the rat brain myelin proteolipids apoproteins.

High variability and low recoveries are obtained when conventional ethanol precipitation is used to recover RNA from the 70-90% formamide-containing solutions that are used to elute poly(U)-Sepharose columns [452]. Precipitation of RNA at different formamide concentrations shows that the variability of the recovery increases with increasing formamide concentration and that the recovery decreases by 7% for each 10% increase in the formamide concentration. The formamide concentrations curve shows that these factors can be optimized by diluting the formamide-containing solutions to at least 30% formamide prior to precipitation.

A micromethod for the determination of the electrophoretic profile of the various poly(A) RNA species in an RNA sample consists of the following steps. The molecules in an RNA sample are first separated according to their molecular weight by electrophoresis in agarose at low ionic strength. The molecules thus separated are then submitted to a second electrophoresis in "binding buffer" in a direction perpendicular to the first one. In the course of this electrophoresis the poly(A) RNA species are separated from other RNA species as they bind to a poly(U)-containing glass-fiber filter which is placed across the electrophoresis path. The method was used to determine the electrophoretic profile of *Rhynchosciara* salivary gland mRNA [201] and to investigate two identical RNA samples (e.g., split samples of total RNA from salivary glands of *Chironomus tentans*) [157]. Poly(A) RNA can be eluted from the poly(U) filter with formamide and subjected to electrophoresis without subsequent precipitation in ethanol. No measurable quantities of rRNA or tRNA are retained on the poly(U) glass-fiber filters. The hybridization technique enables a quantitative retention of poly(A) molecules representing a wide range of chain lengths.

The isolation of structural genes whose transcripts do not contain terminal poly(A) sequences has been described [158]. Poly(A) tails of predicted length were synthesized at the 3'-OH ends of RNA molecules employing *E. coli* ATP:RNA adenyltransferase. The gene isolation was performed in two steps: (a) enrichment of DNA fragments carrying the genes of interest, and (b) subsequent cloning and amplification of these fragments. The enrichment of given gene sequences requires separate purifications of each of two complementary DNA chains, followed by their annealing. Rat liver 5S RNA genes were isolated by this method. DNA (+) chains were obtained by hybridization EcoRI digested and denatured rat liver DNA fragments with the 5S poly(A) RNA. The complex was purified on an oligo(dT)-cellulose column. DNA (-) chains were purified by hybridizing DNA fragments with the 5S cDNA, which was coupled to oligo(dT)-cellulose. In the isolated and annealed double-stranded DNA preparation, the enrichment of 5S DNA was 350- to 400-fold compared with the total genomic DNA. These purified DNA sequences were inserted into the EcoRI site of pBR322. Of 1200 recombinant-plasmid colonies analyzed in the colony hybridization test, 4 hybridized with [^{125}I]5S RNA.

Copia-specific RNA was isolated from *Drosophila melanogaster* tissue culture cells by hybridization of cytoplasmic poly(A) RNA to copia DNA immobilized on cellulose [159]. The purified RNA was translated in reticulocyte lysates.

Antibodies specific to double-stranded RNAs were purified from complexes of poly(A)·poly(U) and anti-poly(I)·poly(C) antibodies by a method of dissociation with 5% dimethylsulfoxide-alkali solution and DEAE-cellulose column chromatography. The purified antibodies were coupled to CNBr-activated cellulose [465]. Synthetic and naturally occurring double-stranded RNAs were bound specifically by the antibody-linked cellulose and were eluted with alkali solution (pH 10.5) or 10% dimethylsulfoxide

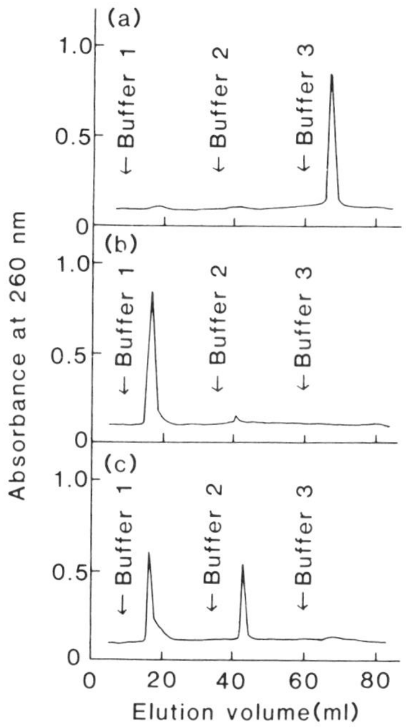

Figure 8 Chromatography of nucleic acids on antipoly(I)·poly(C) antibody-linked cellulose. The column (1.0 × 3.2 cm), which contained 2.8 mg of antibodies coupled to 0.6 g of cellulose, was packed and equilibrated with 0.01 M Tris-HCl (pH 7.6) containing 0.15 M NaCl (buffer 1). Various nucleic acids (100 μg) were eluted stepwise from the column at a flow rate of 40 ml/hr by 0.1 M Tris-HCl (pH 7.6)/0.15 M NaCl (buffer 1); 0.01 M Tris-HCl (pH 7.6) containing 1 M NaCl (buffer 2); and 0.02 M sodium bicarbonate (pH 10.5) (buffer 3). (a) Poly(I)·poly(C); (b) baker's yeast tRNA; (c) *E. coli* rRNA. (From Ref. 465.)

(Fig. 8). The antibody-linked cellulose column appeared to be of practical value for the separation of rice dwarf virus ribonucleic acids from rice plant nucleic acids (Fig. 9).

Cytoplasmic and nuclear RNA is prepared from chicken embryo skin, muscle, and brain at 13 days of incubation. The concentration and labeling of RNA is measured after oligo(dT)-cellulose chromatography [160].

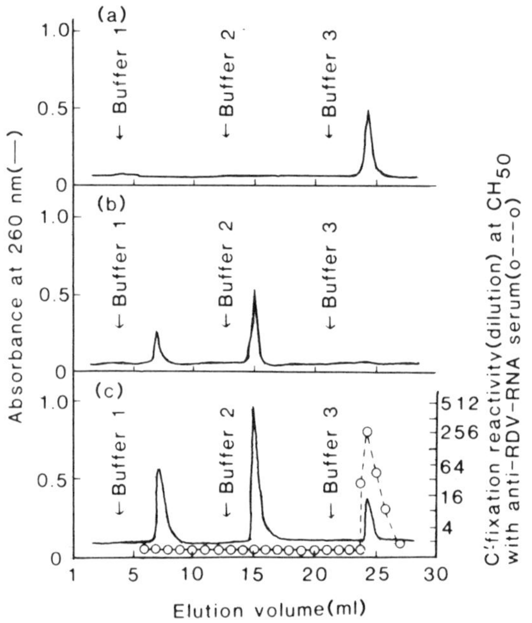

Figure 9 Isolation of rice dwarf virus RNA from rice plant nucleic acids by anti-poly(I)·poly(C) antibody-linked cellulose. The samples were applied on the column (1.0 × 3.6 cm, 0.7 g of cellulose bound 2.24 mg of antibodies) and eluted in the manner described in the legend of Fig. 8. (a) Rice dwarf virus RNA (100 μg); (b) rice plant nucleic acids (100 μg); (c) rice dwarf virus RNA (50 μg) plus rice plant nucleic acid (500 μg). (From Ref. 465.)

m-Aminophenylboronate-substituted agarose binds specifically RNA chains carrying a mature 5' cap [463]. The binding occurs most effectively at pH > 8 and involves diester formation between the negatively charged tetrahedric boronate group and the cis-diol of the ribose of the cap. The positive charge introduced by the m^7G methylation is necessary for efficient binding, although two closely spaced cis-diol groups alone are sufficient for binding. Noncapped RNA [like poly(U) and rRNA] and decapped RNA are not bound. It is shown that the matrix can be used for the isolation of capped small nuclear RNA and mRNA.

A highly charged component can be isolated from a total RNase T2 digest of nuclear polyadenylylated RNA from HeLa cells that is separable from caps by (dihydroxyboryl)aminoethyl cellulose chromatography [478]. Chemical and enzymatic analyses show that the component contains a 2'-5' phosphodiester bond that creates a branch at the 2'-hydroxyl group of one nucleotide already linked to an adjoining nucleotide through the usual 3'-5' phosphodiester bond. The structure was confirmed by analysis of a similar component isolated from nuclease PI digests of the same nuclear polyadenylylated RNA. Branches occur in roughly 10% of nuclear polyadenylylated RNAs, including those greater than 10S in size, but are absent from cytoplasmic polyadenylylated RNA.

II. SELECTED EXAMPLES OF THE RNA ISOLATION

A. Virus RNA

T_4-specific RNA has been quantitatively separated from *E. coli* RNA by chromatography on columns made of T_4 DNA linked to cellulose [92]. The fraction of poly(A) nuclear RNA from SV40 transformed mouse cells contain a high percentage of virus-specific sequences [161]. Large primary transcripts containing host and virus-specific sequences were analyzed by means of localization of virus sequences on them. The molecules of heavy (>28S) nuclear poly(A) RNA were fragmented and 3'-poly(A) fragments were separated by poly(U)-Sepharose. These fragments were then hybridized with SV40 DNA. It was shown that 3'-poly(A) fragments of pre-mRNA were enriched in virus-specific sequences in comparison with complete molecules.

The isolation of RNA fragments originating from defined regions of phage Qβ RNA minus strands is described [162]. Large RNase T_1 oligonucleotides were isolated on a preparative scale from Qβ RNA. After addition of AMP residues (50, on average) using terminal adenylate transferase, these pure oligonucleotides were hybridized to [^{32}P]-labeled Qβ RNA minus strands synthesized in vitro. Fragments in the size range 100 to 500 nucleotides were then generated by partial digestion with RNase T_1. Fragments hybridized to such oligonucleotides were recovered by chromatography on poly(U)-Sephadex and then resolved according to their size by polyacrylamide gel electrophoresis. Adenovirus 2 and vaccinia RNA were specifically bound to matrices containing the homologous A_{20} DNA [85].

Sendai virus synthesized, in vitro, [^{32}P]AMP- and [^{3}H]AMP-labeled RNA that ranged in size from 3S to 25S with major peaks at 7S and 13S [163]. Both labeled products were predominantly single-stranded RNA and were complementary in base sequence to 50S virion RNA. Passage of the 3S to 25S in vitro RNA transcripts through a poly(U)-cellulose column revealed that only the larger (predominantly 18S) RNA transcripts contained poly(A) segments capable of binding to the column.

Of the 10 subgenomic-sized RNAs of tobacco etch virus detected, 4 were found in both virion RNA and infected tissue RNA, 3 others only in virion RNA, and 3 only in infected tissue RNA [468]. Of the last three, two may be engendered during isolation by nuclease action on virion RNA as indicated by controls. On the basis of their binding to oligo(dT)-cellulose, all of the RNAs found in virus-infected tissue appeared to be polyadenylated.

Chromatography on poly(U)-cellulose and sedimentation and electrophoresis analysis were used to study poly(A) virus-specific RNA of Newcastle disease virus [164]. The poly(A)-containing molecules were shown to be present in approximately similar amounts (60-80%) in 18S, 24S, and 35S RNA.

The 34S RNA subunits of three RNA tumor viruses (virus from spontaneously activated AKR cells, Moloney leukemia virus, and RD114 virus) were examined for differences in poly(A) content [165]. Poly(U)-Sepharose columns and a thermal elution were used to fractionate the subunits. The results indicate that the subunits are heterogeneous with respect to their poly(A) contents and that there may be significant differences among RNA viruses with respect to the size and distribution of their poly(A) regions.

RNA of tomato spotted wilt virus (TSWV) was analyzed by gel electrophoresis on methyl mercury-agarose gels [129]. Three segments were detected. The behavior on oligo(dT)-cellulose columns indicated that TSWV RNA does not contain poly(A) sequences. TSWV RNA directed the synthesis of several proteins in a wheat germ cell-free system and a messenger-dependent rabbit reticulocyte lysate. Analysis of the translation products revealed that two major polypeptides were synthesized in both systems. One would be identified by immunoprecipitation as the nucleocapsid protein; the identity of the other is unknown. The results demonstrate that TSWV contains a positive-stranded RNA. TSWV RNA labeled with ^{125}I did not hybridize with polyribosomal RNA from infected leaves, and transcriptase activity could not be detected in purified preparations of TSWV or in nucleocapsid extracts from infected leaves. This study supports the view that TSWV is a unique membrane-bounded plant virus whose genome consists of three RNA segments which act as mRNA.

Media from cells producing RNA tumor viruses, when treated with sodium dodecyl sulfate and poly(vinyl sulfate), yield 70S RNA as the major species binding oligo(dT)-cellulose [166]. The procedure described is suitable for rapidly processing large quantities of media for purifying RNA from commercially available virus, with a 5- to 10-fold higher yield than was obtained using existing methods.

Influenza viral complementary RNA was purified free from any detectable virion-type RNA, and its genetic content and activity in wheat germ cell-free extracts were examined [167]. Fragments of foot-and-mouth disease virus RNA of decreasing size containing the 3'-poly(A) sequence have been prepared by alkali treatment and sucrose gradient centrifugation followed by oligo(dT)-cellulose chromatography [168]. Polyacrylamide gel electrophoresis of the RNase T_1-resistant oligonucleotides from these poly(A) fragments has enabled the location of some of the longer oligonucleotides on the RNA. The poly(C) tracts has been located about 400 nucleotides from the 5' end of the RNA by analysis of the products from the digestion of the RNA with RNase H in the presence of oligo(dG) [169].

By chromatography on poly(A)-Sepharose column, it has been shown that poly(U) sequences exist on the negative strands on Sindbis virus-specific RNAs [170]. The isolated RNAs are negative- and positive-strand hybrids.

Sindbis virus 26S [^{3}H]-labeled RNA was mixed with [^{32}P]-labeled vesicular stomatitis virus (VSV) poly(A) mRNA and loaded onto a column of Sindbis cDNA-cellulose [171]. After extensive washing at 30°C, 37% of the input [^{3}H]- and 47% of the [^{32}P]-labeled RNAs were retained. When the column temperature was raised to 60°C, all the [^{32}P]-labeled VSV RNA was eluted. In contrast, 27% of the input [^{3}H]-labeled Sindbis RNA remained bound and was released when the elution temperature was increased to 90°C.

Uninfected mouse kidney cells and mouse leukemia cells L1210 in culture contained a series of 4.5S RNAs which was structurally identical to the series of 4.5S RNAs associated with genomic RNAs of murine retroviruses and poly(A) RNAs from virus-infected cell [172]. Normal rat kidney cells and baby hamster kidney cells in culture also contained a series of 4.5S RNAs. The structures of the 4.5S RNAs from mouse, rat, and hamster cells were very similar, but not identical. These 4.5S RNAs were not found in cultured cells of other vertebrates, such as human, monkey, cat, mink, rabbit, and chicken cells.

A series of low-molecular-weight RNAs (4.5S-5.5S) as well as other 4S-7S RNAs were dissociated from genomic RNA of spleen focus, forming virus (SFFV) by heating [466]. On two-dimensional polyacrylamide gel electrophoresis, this series of RNAs gave a series of more than 30 spots. RNase T1 fingerprints of these spots were identical except for differences in 3'-terminal oligonucleotides, which were due mainly to different numbers of uridylic acid residues, larger RNA molecules containing poly(U) sequences at their 3' termini. This series of RNAs is also associated with poly(A)-containing nuclear and cytoplasmic RNAs from SFFV-infected cells. Poly(U) containing RNA was purified on poly(A)-Sepharose.

70S RNA has been prepared from mouse mammary tumor virus (MMTV) produced by the GR tumor cell line [173]. After denaturation the RNA was applied to a sucrose gradient and molecules sedimenting between 10S and 16S were selected and passed over an oligo(dT)-cellulose column. The poly(A) RNA fraction was used as a template for the synthesis of cDNA.

The RNA of a type D retrovirus (PMFV) isolated from and propagated in a human cell line has been characterized [467]. Like other retroviruses, PMFV contains a 60S-70S RNA. When denatured by heat, the high-molecular-weight RNA prepared from fresh virus yielded subunits of about 30S. As shown by poly(U)-Sepharose chromatography, the 60S-70S RNA as well as the 30S subunits contain a poly(A) stretch of approximately 80-200 nucleotides.

B. Bacterial RNA

Pulse-labeled RNA isolated from *E. coli* cells was analyzed by oligo(dT)-cellulose chromatography and by Millipore binding assay for poly(A) RNA [174]. Whereas poly(A) RNA amounted to as much as 15% of the total pulse-labeled RNA from cells grown on limiting phosphate medium, pulse-labeled RNA from cells grown on phosphate medium gave values around 1-5%. Steady-state labeled RNA from cells grown on limiting phosphate medium contained 1-2% poly(A) RNA. The addition of poly(A) sequences appears to be post-transcriptional. These results strongly favor the view that bacterial mRNAs may contain poly(A) stretches.

The appearance of a population of poly(A) RNA during sporulation of *Bacillus subtilis* has been reported [175]. This poly(A) RNA was characterized by hybridization to [^{3}H]poly(U), by its ability to bind oligo(dT)-cellulose, and by its resistance to RNase A and T_1. Most of the poly(A) sequences are located at the 3' terminus of the RNA chains and are about 160-180 nucleotides in length.

Chromatography of pulse-labeled RNA from *B. subtilis* on oligo(dT)-cellulose or poly(U)-agarose revealed the presence of 10-20% poly(A) RNA [177]. The poly(A) content of total *B. subtilis* RNA was found to be 0.091%, indicating that the steady-state level of poly(A) RNA in *B. subtilis* is 2.3% of the total RNA or approximately 25% of the mRNA. The mRNA activity of poly(A) RNA and of total *B. subtilis* RNA was compared by using a cell-free translation system from *E. coli*. Poly(A) RNA was eight times more effective than total RNA in stimulating protein synthesis, indicating a considerable enrichment of mRNA in this fraction. The sequence complexity of poly(A) RNA was examined by hybridization with restriction endonuclease-generated fragments of *B. subtilis* DNA, and complementarity to many different DNA sequences was found.

A one-step procedure for the isolation of bacterial RNA, involving lysis by proteinase K in the presence of sodium dodecyl sulfate, has been described [176]. Pulse-labeled RNA isolated by this procedure from *Bacillus brevis*, *B. subtilis*, and *E. coli* B has been found to contain a substantial fraction (15-40%) of poly(A) RNA as determined by adsorption to oligo(dT)-cellulose. This contrasts with RNA isolated by procedures involving phenol extraction, a process that appears to lead to the selective loss of poly(A) RNA. The presence of poly(A) RNA in *E. coli* was confirmed by hybridization with [^{3}H]poly(U) acid. Using the proteinase K method for RNA isolation, it was possible to demonstrate the in vitro synthesis of poly(A) RNA by toluene-treated cells of *B. brevis*, *B. subtilis*, and *E. coli*.

Up to about 50% of the total radioactivity in pulse-labeled RNA in *B. brevis* 47-5 was found in the polyadenylated fraction isolated by adsorption to oligo(dT)-cellulose [469]. Labeled RNA was bound to the cellulose regardless of whether the radioactive precursor was [^{3}H]adenosine or [^{3}H]uridine, showing that the adsorbed material was poly(A) RNA rather than free poly(A). Poly(A) tracts, isolated after digestion of pulse-labeled RNA with pancreatic and T_1 RNases, were homogeneous, with a length of about 95 nucleotides. Susceptibility of the isolated poly(A) tracts to degradation by snake venom phosphodiesterase and polynucleotide phosphorylase indicated that the poly(A) sequences were located directly at the 3' terminal of the RNA molecules. An improved method for isolating highly active polysomes from *B. brevis*, which contained poly(A) RNA in a polydisperse distribution, has been reported [510]. The authors proposed that this poly(A)

RNA is mRNA. They present evidence that the amount of poly(A) RNA in the exponential phase increases toward the transient phase of growth (i.e., the deceleration phase between exponential growth and the stationary phase). Coding sequences present in the poly(A) RNA from *B. brevis* are examined by in vitro translation in a highly efficient cell-free system. Newly synthesized polypeptides analyzed on autoradiograms indicate that poly(A) RNA is indeed mRNA and that polypeptides encoded by poly(A) RNA seem to be identical with those encoded by poly(A)-deficient RNA.

C. Plant RNA

Cell suspension cultures from parsley were labeled in vivo with [2-^{3}H]adenosine. The RNA isolated from the ribosomal pellet was fractionated on an oligo(dT)-cellulose column [178]. Approximately 1.5% of the RNA was retained at high salt concentrations and eluted at low ionic strength. This fraction contained poly(A) segments with an average length of about 80 nucleotides. It was active as template in a cell-free system from wheat germ.

Poly(A) RNA from *Vicia faba* meristematic root cells was fractionated on the basis of mean poly(A) size by a thermal stepwise elution from poly(U)-Sepharose [179,180]. Such a procedure allowed the elimination of contaminating RNA at 30°C and the collection of two populations of purified poly(A) RNA at 40°C and 50°C, respectively. RNA eluting at the higher temperature carried a poly(A) segment (mean size of 100 nucleotides) twice as large as the RNA eluting at the lower temperature.

Total polysomal RNA from yellow lupin root nodules was fractionated by double oligo(dT)-cellulose chromatography [132]. Poly(A)$^+$ and poly(A)$^-$ RNA fractions showed considerable messenger activity in wheat germ and rabbit reticulocyte cell-free systems. The sizing of poly(A) RNA on sucrose density gradient gives rise to separation of 14S mRNA from 22S-24S mRNA species. A single polypeptide with molecular weight 22,000 was coded for by 14S mRNA, while two polypeptides were the main products of 22S-24S mRNA fraction.

Poly(A) RNA was isolated from maize by chromatography on oligo(dT)-cellulose [181]. A modified nucleotide ("X") was detected in RNase T_2 digests of the RNA as part of a resistant dinucleotide. Intact poly(A) tracts were released from poly(A) RNA by digestion with RNase T_1 and A in a high-salt buffer and were isolated by oligo(dT)-cellulose chromatography, the poly(A) fragments varying in chain length from approximately 17 to 70. The modified nucleotide was shown to occupy an internal position in these poly(A) tracts.

Poly(A)$^+$ and poly(A)$^-$ RNA from wounded potato tuber tissues and crown gall tumors were separated from total RNA by oligo-(dT)-cellulose chromatography [131]. The poly(A)$^+$ RNA was characterized by sucrose density gradient centrifugation, hybridization with [^{3}H]polyuridylic acid, and in vitro translation in a rabbit reticulocyte lysate system. The tumor poly(A)$^+$ RNA was a heterodisperse mixture. Upon poly(U) hybridization of the gradient fractions, two major hybridization peaks at 7S and 21S and two peaks at 11S and 16S appeared. In an in vitro translation system the poly(A)$^+$ RNA programmed the synthesis of 23 different polypeptides. The 21S poly(A)$^+$ RNA was about five times more active than the 7S poly(A)$^+$ RNA in in vitro protein synthesis. The poly(A)$^+$ RNA from wounded tissues was also heterodisperse. This RNA coded for at least 28 polypeptides, which were different from those of crown gall tumor tissues. The translation of tumor poly(A)$^+$ RNA was completely blocked by 0.5 mM 7-methylguanosine-5'-phosphate, but not by 7-methylguanosine, suggesting the presence of a 5'-cap structure.

In potato bud tissue, after gamma irradiation there was a two- to three-fold increase in RNA synthesis which was sensitive to actinomycin D. The poly(A) RNA was separated from the total RNA synthesized during this phase by chromatography on poly-(U)-Sepharose [470]. About 1% of the total RNA synthesized was polyadenylated. Poly(A)$^-$ RNA was separated into 11 bands on electrophoresis and a major portion of the radioactivity was found in four bands.

D. Heterogeneous Nuclear RNA

Three classes of HnRNA, α, β, and γ, can be identified in the early sea urchin embryo by hybridization with poly(U) [182]. The relative amounts of these classes vary as a function of embryonic development. It is demonstrated that the poly(A)-containing classes of HnRNA, α and β, can be subfractionated by hybridization with poly(A)-agarose into species containing and lacking oligo(U) segments. These oligo(U) segments could not be detected in γ HnRNA, which was previously shown also to lack adenylate segments. The relative proportions of these species undergo marked changes during development from early blastula (7 hr) to mesenchyme blastula (20 hr).

Poly(A) sequences have been found covalently linked in heterogeneous DNA-like nuclear RNA of HeLa cells [203]. This poly(A) material seems homogeneous in size and accounts for about 0.5% of such RNA. Similar poly(A) sequences were found in rapidly labeled polyribosomal RNA, thought to be mRNA. Poly(A) segments derived from either HnRNA or mRNA molecules with T1 RNase consisted of two uridylate and one to two cytidylate residues per 195 adenylate residues and were devoid of any GMP residues [204]. Pancreatic RNase produced poly(A) segments approximately 195 nucleotides long which consisted exclusively of adenylic acid. These experiments indicated that at least 80-90% of the poly(A) in HnRNA and mRNA is located at the 3'-OH termini and that the nucleotides adjacent to the poly(A) in HnRNA and mRNA are the same. This result further supports the hypothesis that some HnRNA molecules are precursors to mRNA molecules. HeLa cell HnRNA derived from high-molecular-weight nuclear ribonucleoprotein (RNP) particles contains oligo(U) sequences of 15-50 nucleotides base-paired with poly(A) [183]. These duplexes are resistant to pancreatic RNase at 0.5 NaCl in native RNP, and remain so after chemical deproteinization of the RNP digests and then copurity with poly(A) on oligo(dT)-cellulose chromatography. Oligo(dT)-cellulose binding capacity of the oligo(U)·poly(A) duplexes is abolished by prior titration of the nonduplex poly(A) regions with excess poly(U). The oligo(dT)-purified fraction is 97.5 mol % A + U and the [^{3}H]uridine-labeled component is resistant to redigestion by pancreatic RNase at 0.5 NaCl but not at 0.01 M NaCl. After thermal denaturation, the [^{3}H]uridine-labeled chains become RNase sensitive at 0.5 M NaCl. Electrophoresis of [^{3}H]adenosine- or [^{3}H]uridine-labeled material in polyacrylamide gels containing 99% formamide confirms that the oligo(U) sequences are not covalently linked to poly(A). The oligo(U)·poly(A) duplexes appear to be associated with protein in native heterogenous nuclear RNP, as reflected by the differential pancreatic RNase sensitivity of the duplexed oligo(U) in RNP (resistant) and RNA (sensitive), measured at physiological ionic strength.

The presence in nuclei and cytoplasm of cultured BHK-21/C13 cells (baby hamster fibroblasts) of RNA species with high affinity for poly(A) was detected using either a Millipore-poly(A)-binding assay or columns of poly(A)-Sepharose 4B [184]. The nuclear species that labels rapidly is metabolically unstable and appears to be a specific subclass of HnRNA. Digestion with T_1 RNase gives rise to a low level (1%) of oligonucleotide fragments of a disperse size range which are relatively rich in uridylate residues. The cytoplasmic species with affinity for poly(A) is similar in size to poly(A) mRNAs and also associates with polysomes. However, it appears to be distinct from the poly(A) mRNAs by virtue of an unusual base composition and relative metabolic instability.

The results of the isolation of oligo(U) containing HnRNA from Friend erythroleukemic cells established that rodent cell HnRNA contain oligo(U) sequences and demonstrate that HnRNA containing both a poly(A) and an oligo(U) sequence can be separated from other classes of HnRNA [474]. Chick embryonic RNA was fractionated by chromatography on oligo(dT)-cellulose and poly-(U)-Sepharose [475] into three classes: poly$(A)^+$ RNA containing poly(A) segments of 100 and more residues, poly$(A)^-$ oligo-$(A)^+$ RNA containing oligo(A) segments of about 25 residues, and poly$(A)^-$ oligo$(A)^-$ RNA which bound to neither of the beds used and which contained double-stranded segments of 300 and more base pairs. These three classes of RNA were found in cytoplasmic as well as in heterogeneous nuclear RNA. Double-stranded segments in HnRNA, unlike those in cytoplasmic RNA, were intermolecular in nature; this may explain the occurrence of "giant" molecules in HnRNA.

E. RNA of Mitochondrial Fraction

Total RNA was extracted from purified mitochondrial and cytoplasmic fractions of germinating conidia of *Trichoderma viride* and bound to oligo(dT)-cellulose at 22°C and 4°C [185]. Under chromatographic conditions which retained very short poly(A) segments (i.e., 4°C), up to 10% of short-term [^{32}P]-labeled RNA from the mitochondrial fraction was selectively bound.

In contrast to other findings, a poly(A) RNA has been isolated from mitochondria of *S. cerevisiae* using poly(U)-Sepharose chromatography [186]. This RNA contains a short poly(A) segment of 20-30 nucleotides and the small size of this segment may account for the failure of others to detect yeast mitochondrial poly(A).

RNA that contains poly(A) has been isolated from yeast mitochondria by poly(U)-Sepharose 4B column chromatography [187]. Pulse-labeled poly(A) RNA shows 8-10 discrete peaks by acrylamide gel electrophoresis. The specific activity of mitochondrial poly(A) RNA is six to eight times greater than that of mitochondrial rRNA after pulse labeling of protoplasts with [^{3}H]uridine. Ethidium bromide inhibits incorporation by over 90%. The total mitochondrial RNA preparation was contaminated with 5-15% cytoplasmic rRNA as determined by gel electrophoresis, but RNA exhaustion hybridization experiments indicated little or no cytoplasmic contamination of the mitochondrial poly(A) RNA. The poly(A)

RNA stimulates [^{3}H]leucine incorporation into protein in an *E. coli* cell-free system. A fraction of the labeled product is precipitated with antibody directed toward yeast cytochrome oxidase, but not with antibody directed toward bovine serum albumin. Sodium dodecyl sulfate-gel electrophoresis of the immunoprecipitated material reveals labeled peptides having the mobility of the three larger cytochrome oxidase peptides, which are known to be translated by mitochondrial ribosomes.

Rat liver mitochondria are able to incorporate labeled nucleotides into RNA species which are retained by Millipore filters or an oligo(dT)-cellulose column [188]. The radioactivity retained was completely RNase sensitive when [^{3}H]UTP was used as radioactive precursor, whereas by using [^{3}H]ATP this fraction was only 30% RNase sensitive. These results demonstrate that rat liver mitochondria are able to synthesize poly(A) RNA species and furthermore that the polyadenylation reaction occurs within the organelle.

F. Synthesis and Degradation of Poly(A) RNA

When the RNA from both the nucleus and polysomes of germinated cotton seeds were fractionated on poly(U)-cellulose columns it was found that both contained poly(A) RNA [189]. In ungerminated seeds most of the poly(A) RNA is located in the nucleus of the cell. 3'-Deoxyadenosine was found to inhibit the synthesis of both nuclear and polysomal poly(A) by 75% in germinated seeds, but had little or no effect on protein synthesis during the first 6 hr of germination. After this time 3'-deoxyadenosine had a pronounced inhibitory effect on protein synthesis, presumably due to the inhibition of poly(A) addition to newly synthesized mRNA.

Mechanical wounding of potato tubers induced a rapid synthesis of RNA in the wounded tissues [130]. Both total and poly(A) RNA increased with time after wounding. Treatment of wounded tissues with the synthetic hormone 2,4-dichlorophenoxyacetic acid further stimulated their syntheses. The poly(A)$^+$ RNA from hormone-treated tissues was more active in in vitro protein synthesis. The in vitro translation of poly(A)$^+$ RNA from both hormone-treated and untreated tissues was inhibited by 7-methylguanosine-5'phosphate, while 7-methylguanosine had no effect, suggesting that both poly(A)$^+$ RNAs contained a blocked 5'-cap structure and that the cap structure was important for in vitro protein synthesis.

The action of cordycepin on nascent nuclear RNA and poly(A) synthesis in regenerating liver has been investigated [190].

Fractionation of HnRNA on poly(U)-Sepharose following 20 mg per kilogram of cordycepin revealed that a 65% reduction occurred in the labeling of poly(A) HnRNA, while nonpoly(A) HnRNA was reduced by 43%. Studies of the pool size and specific radioactivity of UTP in control and cordycepin-treated animals indicated no significant alterations in these parameters. Assessment of poly(A) size using poly(A) HnRNA annealed with oligo(dT)$_{10}$ as template primer for *E. coli* DNA polymerase I showed that 20 mg per kilogram of cordycepin inhibited nuclear polyadenylation by 43%; no alteration in the binding of poly(A) HnRNA to Millipore filters occurred at this dose of cordycepin. These results indicate that cordycepin is a nonselective inhibitor of nuclear RNA and poly(A) synthesis in regenerating rat liver.

RNA synthesized during the first 1 hr of conidial germination has been characterized by sucrose-sodium dodecyl sulfate gradient centrifugation, binding to poly(U) filters, and oligo(dT)-cellulose chromatography [191]. At all labeling periods examined, poly(A) RNA is synthesized and associated with polysomes. In addition, ribosomal RNA is synthesized, processed, and incorporated into polysomes. Approximately 40% of the labeled RNA sedimenting between 5S and 17S binds to poly(U) filters. RNA which binds to oligo(dT)-cellulose displays a heterogeneous distribution in sucrose-sodium dodecyl sulfate gradients with a major, broad peak at 10S-16S. In addition, some poly(A) RNA sediments beyond the 25S marker. Approximately 3% of the [^{3}H]adenosine in pulse-labeled polysomal RNA is in poly(A) segments resistant to pancreatic and T_1 RNase.

Bacillus emersonii zoospores and germlings were assayed for the presence of poly(A) RNA by the poly(U)-glass-fiber filter binding technique and oligo(dT)-cellulose chromatography [192]. Results show that poly(A) RNA is synthesized de novo during germination of zoospores. In addition, a portion of the preformed mRNA in the zoospore has a poly(A) fragment attached to the 3' terminus. During germination some of the preformed mRNA is adenylated. Experiments with actinomycin D demonstrated that transcription is not required for polyadenylation of zoospore mRNA. These findings indicate that lack of a poly(A) fragment is not an explanation for the inability of the zoospore messages to participate in protein synthesis.

The effect of estrogen on the sequence and population complexity of chick oviduct poly(A) RNA has been investigated. Total cellular RNA preparations were isolated from chicken oviducts at three different development stages which were then isolated from these preparations using oligo(dT)-cellulose chromatography [193].

The number average nucleotide length of the poly(A) RNA preparations in each case was approximately 2000 nucleotides. The number average nucleotide length of the poly(A) residues at the 3'-terminal end of each RNA preparation was approximately 70 adenylate residues. cDNA copies to each preparation of poly(A) RNA were synthesized. These cDNA poly(A) preparations were then utilized in DNA excess hybridization experiments to analyze the complexity of the DNA sequences from which these RNAs were transcribed. Approximately 22% of each of the total cellular poly(A) RNAs were transcribed from repeated DNA sequences while the remaining majority were transcribed from single-copy or unique-sequence DNA. It was possible to estimate the number of different poly(A) RNA sequences per cell.

During late lytic infection of mouse kidney cell cultures polyoma 16S and 19S (late 19S RNA) were isolated by oligo(dT)-cellulose chromatography [194,195]. Approximately 60-80% of total cytoplasmic polyoma RNA contained tracts of poly(A). Early in lytic infection when viral DNA synthesis and the production of capsid protein are blocked by the addition of 5-fluorodeoxyuridine, approximately 100% of polyoma "early" 19S RNA was quantitatively retained by oligo(dT)-cellulose, indicating the presence of poly(A) tracts on most 19S mRNA molecules. In addition, two classes of polyoma RNA, synthesized after the onset of cellular RNA synthesis under conditions where DNA synthesis is inhibited with 5-fluorodeoxyuridine, were found to contain tracts of poly(A). These species, sedimenting at 16S and 19S in aqueous sucrose density gradients, were also quantitatively retained by oligo(dT)-cellulose.

A portion of rat brain RNA contains poly(A) sequences [196]. Young rats have a greater amount of brain RNA which contains poly(A) than do adult animals. The length of the poly(A) sequence from the brain RNA of young animals was shown to be somewhat longer than that from the RNA of adults. The RNA that bound to the oligo(dT)-cellulose was found to be large and heterogeneous, and to be almost free of ribosomal or of low-molecular-weight RNAs, to priming a cell-free protein-synthesizing system. The RNA that bound to the oligo(dT)-cellulose was much more active than the unbound material to priming a cell-free protein-synthesizing system. However, in the case of the nuclear RNA there was not as great a difference between the material that was bound and that which was not bound.

Synaptosomal RNA of rat brain was labeled in vivo by intracranial injection of tritiated uridine. The change in the specific activity of this material with time was similar to that of polysomal

RNA. The percent of the radioactive synaptosomal RNA that bound to oligo(dT)-cellulose columns [197] decreased with time after intracranial labeling. The percent of the total synaptosomal RNA that bound to oligo(dT)-cellulose was greater than that of polysomes. The length of the poly(A) sequence of synaptosomal RNA was approximately one-half that of polysomal RNA and about the same as that from mitochondria. Investigation of synaptosomal RNA using sucrose gradients and polyacrylamide gel electrophoresis indicated that there were several distinct species present, and that they were similar to those from the mitochondria. The poly(A) RNA isolated from synaptosomes stimulated the incorporation of radioactive leucine in a cell-free protein synthesis system. Isolation of RNA from subsynaptosomal components indicated that most, if not all, of the synaptosomal messenger activity was localized in the synaptic mitochondria.

The metabolism of high-molecular-weight RNA in the nuclear and cytoplasmic fractions of newborn and adult rat brain was investigated after intracranial administration of $[^{32}P]P_i$ [198] (Fig. 10). The results suggest that there are considerable changes in the regulation of the nucleocytoplasmic relationship of rRNA and poly(A) RNA during the transition of the brain from a developing replicative phase to an adult differentiated and nondividing state.

The half-life time and the rates of biosynthesis and polyadenylation of nuclear pre-mRNA fractions isolated from the livers and brain cortex of 1-, 7-, 90- and 540-day-old rats and of 21-day-old fetuses by poly(U)-Sepharose chromatography and containing oligo(A) sequences have been studied [476]. It was shown that the rates of biosynthesis and polyadenylation decrease with aging, whereas the half-life time of pre-mRNA increases. The intranuclear maturation of pre-mRNA is increased. The intranuclear maturation of pre-mRNA in both cell types is decelerated.

Nuclear and polysomal brain RNA from the rabbit bind to Millipore filters and oligo(dT)-cellulose, suggesting the presence of poly(A) sequences [199]. The residual polynucleotide produced after RNase digestion of $[^{32}P]$pulse-labeled brain RNA is 95% adenylic acid and 200-250 nucleotides in length. After longer isotope pulses the polysomal poly(A) sequence appears heterodisperse in size and shorter than the nuclear poly(A). Poly(A) sequences of brain RNA are located at the 3'-OH termini, as determined by the periodate-$[^{3}H]NaBH_4$ labeling technique. Cordycepin interferes with the processing of brain mRNA, as it inhibits in vivo poly(A) synthesis by approximately 80% and decreases the appearance of rapidly labeled RNA in polysomes by 45%. Poly(U)-Sepharose chromatography combined with a discontinuous elution

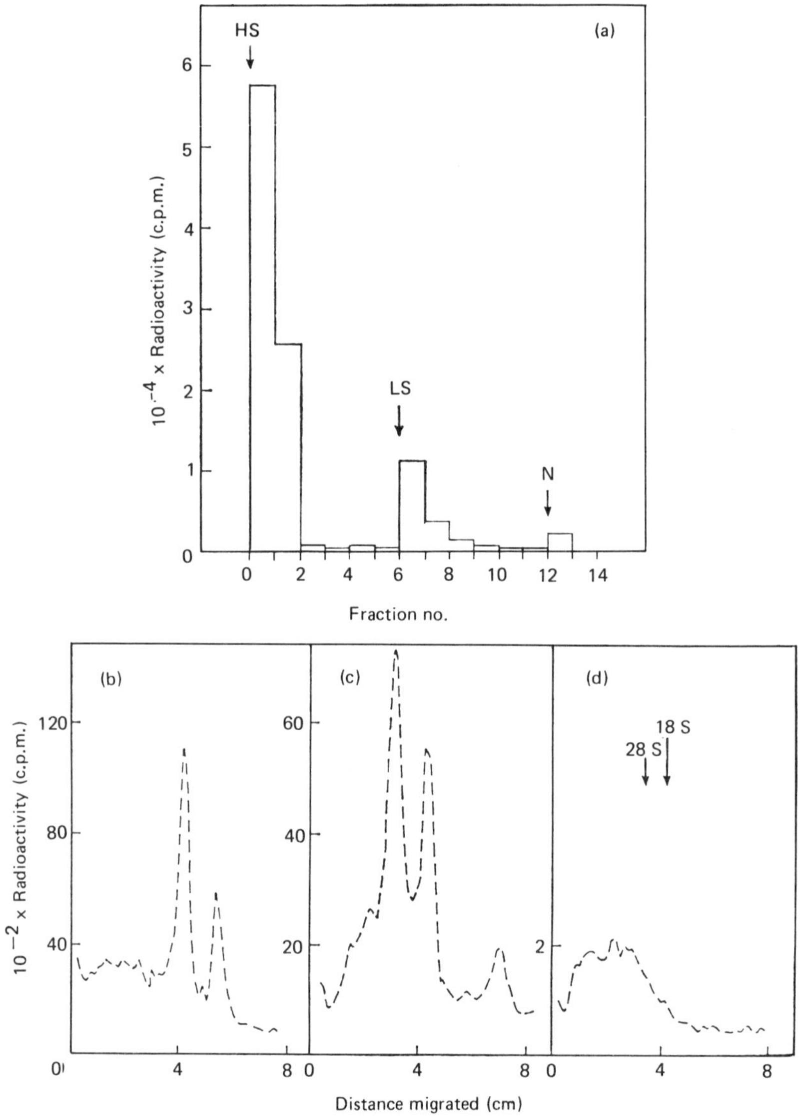

Figure 10 Isolation and characterization of [^{32}P]-labeled poly(A) RNA. (a) High-molecular-weight RNA was isolated from forebrains of 8-day-old rats 5 hr after the intracranial injection of 1 mCi of [^{32}P]P_i per gram of brain. The RNA (80×10^3 cpm in 6 A_{260} units) was applied in high-salt buffer containing 0.4 M NaCl, 10 mM Tris/HCl, pH 7.4, 0.2% sodium dodecyl sulfate, onto a column

scheme is a very sensitive procedure for monitoring metabolic changes in poly(A) RNA subpopulations with time, subcellular location, and RNA extraction procedure [200].

G. Transfer RNA

The chromatographic behavior of yeast tRNA on cellulose derivatives containing nucleoside has been studied (Table 11). Immobilized boronic acid potentially could be used to separate amino

Table 11 Chromatographic Behavior of Yeast tRNA on Cellulose Containing Immobilized Nucleosides

Nucleoside	Moles nucleoside per 100 g of cellulose	Capacity (A_{260} units of tRNA per 100 g of cellulose)	Recovery (%) tRNA (A_{260} units)	Elution range (M NaCl) tRNA at pH 5.5
A	0.1	>1000	100	0.15-0.5
C	0.2	>1000	>80	0.2-0.8
G	0.2	300	100	0.05-0.15
U	0.1	>1000	>80	0.01-0.15
dT	0.2	>1000	90	0.1-0.35
I	0.2	750	>80	0.01-0.2

Source: Ref. 29.

Figure 10 (Continued) of 200 mg of oligo(dT)-cellulose. The RNA not containing poly(A) was eluted in 2-ml fractions with high-salt buffer (HS). Poly(A) RNA was then eluted with 10 mM Tris-HCl, pH 7.4, containing 0.2% sodium dodecyl sulfate (low-salt buffer; LS), again in 2-ml fractions. Residual radioactivity in the column was removed with 0.1 m NaOH (N). Unfractionated high-molecular-weight RNA, (b) RNA not containing poly(A), and (c) poly(A) RNA, (d) were electrophoresed on polyacrylamide gels. Radioactivity measurements are shown. (From Ref. 198.)

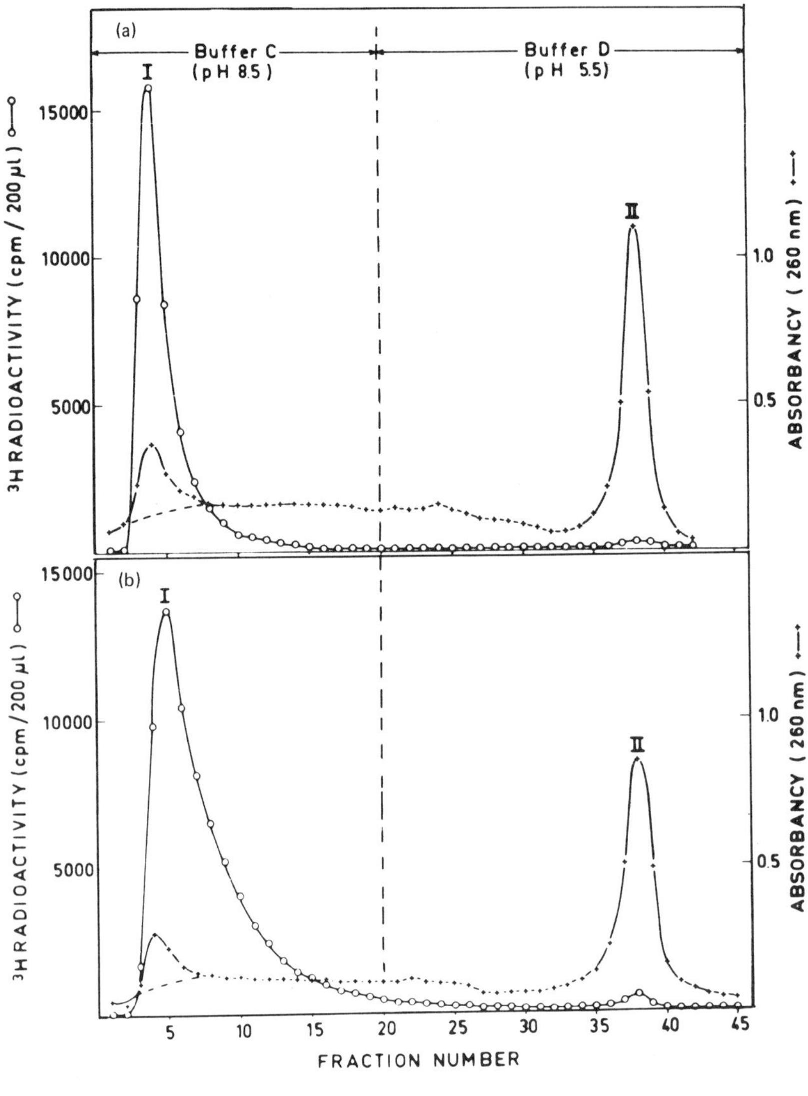

Figure 11 Purification of *E. coli* aminoacyl-tRNA by column chromatography on borate gels. (a) Separation of a tRNA mixture containing 674 pmol of [^{3}H]glutamyl-tRNA (1.6%) in a total of 26.3 A_{260} units of tRNA; (b) separation of a tRNA mixture containing 1030 pmol of [^{3}H]lysyl-tRNA (2.4%) in a total of 27.9 A_{260} units of tRNA. Elution buffers were applied as indicated by the arrows. Fractions of 4.6-ml volume were collected every 5 min in tubes containing 200 µl of 1 M sodium acetate, pH 4.5, in order

acyl-tRNA from unacylated species (Fig. 11) [116,119,123,124]. Highly purified isoaccepting species of tRNA were prepared by use of a polyacrylamide support substituted with nitrobenzene boronic acid as functional groups [125]. tRNA isoacceptors were obtained by enzymatically acylating a mixture of tRNA species with a single amino acid and passing the mixture over the nitrobenzene boronic acid gel at pH 6.5 or 7.0. Pure aminoacyl-tRNA was eluted, and unaceylated tRNA species were bound. The bound species were recovered by lowering the pH of the eluent to 4.5. It is shown that yeast $tRNA^{Phe}$, chemically coupled by its oxidized 3'-CpCpA end, behaves exactly as free $tRNA^{Phe}$ in its ability to form a specific complex with *E. coli* $tRNA_2^{Glu}$ having a complementary anticodon [202].

Elongation factor Tu from *Thermus termophilus* has been coupled to CNBr-Sepharose 4B and its properties investigated [501]. The immobilized elongation factor retained its ligand-binding properties. It specifically binds GDP and GTP but does not interact with other nucleotide 5'-phosphates. A conversion of immobilized EF-Tu·GDP to EF-Tu·GTP can be achieved by simple equilibration with GTP. The immobilized EF-Tu·GTP specifically binds aminoacyl-tRNAs and allows a facile purification of aminoacyl-tRNA species from bulk tRNA. It is stable at room temperature for several months and can be used repeatedly for aminoacyl-tRNA isolation.

Figure 11 (Continued) to neutralize the eluent immediately. Aliquots of 200 μl were assayed for acid-insoluble radioactivity (o). Absorbancy of the individual fractions was followed at 260 nm (+); the absorbancy background of the elution buffer alone (---) is not subtracted in the figure. (Reprinted with permission from Ref. 119. Copyright 1973 American Chemical Society.)

8

Isolation of Messenger RNA

mRNAs are central components in the expression of eukaryotic structural genes. Thus an important objective for many studies on gene regulation is to isolate specific mRNAs for structure-function analysis, and to prepare specific radioactively labeled cDNA to be used as a hybridization probe for the quantitation of mRNA. The cDNA can also be used as an affinity probe for the isolation of specific gene segments and RNA transcription products, and for the determination of nucleotide sequences. mRNA in animal cells appears to be generated by cleavage of high-molecular-weight precursors in the nucleus. These mRNA precursors are contained in the metabolically active pool of RNA molecules (HnRNA). Polysomal mRNA as well as a substantial part of the HnRNA have been shown to contain segments of poly(A) at the 3' terminus of the polynucleotide chains. In addition to this there is a nonpolysomal pool of ribonuclear protein particles in the cytoplasm, which also contains poly(A) RNA. This pool is thought to contain mRNA en route to translation.

I. VARIOUS POLY(A) ADSORPTION METHODS

Poly(A) mRNAs can be separated from other cellular RNAs by adsorption to nitrocellulose filters or by relatively simple methods of chromatography on unmodified cellulose, trityl cellulose/Sepharose, oligo(dT)-cellulose, or poly(U)-agarose/Sepharose. Adsorption techniques employing nitrocellulose filters were often used before the advent of readily available template chromatography supports [205,206]. Although nitrocellulose filters are effective in the isolation of poly(A) RNA, they also nonspecifically adsorb rRNA and small amounts of single-stranded DNA, even after a readsorption step [207]. Lower recoveries of mRNAs from nitrocellulose filtration techniques are also observed, apparently due to a requirement for longer poly(A) segments [208]. Unmodified cellulose has been shown to bind poly(A) [209,210,546-548] at neutral pH in high-salt solution and poly(A) is released in low-salt solution. The salt effect is due principally to the cation, not to the ionic strength of the solution. Monovalent cations are not equally supportive of binding. The order of their effectiveness corresponds to a Hofmeister lyotropic series, increasing with decreasing radius of the hydrated cation. The cellulose-poly(A) interaction does not seem to be due to simple hydrophobic attraction but is consistent with association by π-π interaction between aromatic constituents of lignin of the cellulose and the purine bases of the poly(A). Single-stranded DNA behaves like poly(A) in these respects, but pyrimidine homopolymers do not. Similarly, poly(A)·poly(U) hybrids have no affinity for cellulose preparations. In high-ionic-strength solution, excess poly(U) removes most of the bound poly(A) from the poly(A)-cellulose complex but does not affect the bound single-stranded DNA. Poly(U) can be used to obtain substantial separation of poly(A) RNA from nucleic acid mixtures that contain single-stranded DNA by selectively eluting the poly(A)-containing material from the cellulose-bound mixture of the two. Fractional elution of poly(A) from cellulose can be accomplished by gradually decreasing the salt content of the eluting buffer. The poly(A) oligomers which disengage from cellulose at rather high salt concentrations seem to be shorter than those which are released only at lower salt concentrations. This discrimination may serve as a basis for the fractionation of poly(A) RNAs of which the poly(A) suffixes differ in length. However, mRNA prepared by this technique is usually contaminated with substantial amounts of rRNA.

Template chromatography with columns of oligo(dT)-cellulose has become widely used. The matrix usually consists of oligo(dT) strands about 8-15 nucleotides in length covalently attached to microgranular cellulose.

Poly(U)-Sepharose is another commonly used template chromatography support that requires mRNA-binding conditions similar to those of oligo(dT)-cellulose [211]. The poly(U) ligand is usually somewhat longer than the oligo(dT) ligand and therefore requires stronger elution conditions, such as 70-90% formamide or elevated temperature. Since the poly(U) chain is approximately 100 monomers long and is attached to the matrix at several points, breaks in the chain lead to the loss of only small amounts of polymeric material. Several weeks of storage in suspension at elevated temperatures results in the loss of only 1-2% of the poly(U) and no detectable loss of binding activity for mRNA. The disadvantages of chromatography on poly(U)-Sepharose are the nonquantitative recovery of poly(A)-mRNA and insufficient adsorbent capacity.

Poly(U)-Sepharose and oligo(dT)-cellulose column chromatography have contributed significantly to the rapid progress made in the elucidation of poly(A) RNA structure and function. Conditions that enhance polynucleotide hybridization, such as high-ionic-strength solutions and low temperatures, also enhance poly(A) RNA binding, and conditions that destabilize polynucleotide hybrids, such as low-ionic-strength solutions, high temperature, and formamide, also disrupt the binding of poly(A) RNA to the matrix-bound complementary polynucleotide, and cause elution of the poly(A) RNA from the column.

Several celluloses were tested for their ability to bind homonucleotide oligomers and natural RNAs [212]. The various cellulose preparations were quite different in their ability to bind the oligonucleotides and the RNAs.

The chromatographic behavior of poly(A) homopolynucleotides on different substituted matrices – poly(U); poly(A); phenyl-, octyl-, ethanolamine-, acriflavin-, and DNA-Sepharose; oligo(dT); and MN-cellulose – has been reported [213]. Using different experimental conditions it was possible to evaluate the participation of electrostatic, hydrophobic, hydrogen-bonding, and/or charge-transfer interaction. It was shown that poly(A) interacts nonspecifically with matrices such as acriflavin or DNA-Sepharose as well as with oligo(dT)-cellulose or poly(U)-Sepharose. It was shown that in addition to specific binding of poly(A) sequences with poly(U), the chromatography of poly(A) RNAs on poly(U)-Sepharose is accompanied by nonspecific irreversible adsorption of polynucleotide by Sepharose gel [214]. This disadvantage may be overcome by establishing optimal BrCN/Sepharose ratios during Sepharose activation and by manyfold treatment of poly(U)-Sepharose with ethanolamine immediately before chromatography of RNAs. It was also found that the efficient separation of poly(A^+) RNA preparations from poly(A^-) RNAs is achieved only after double chromatography of RNA on poly(U)-Sepharose.

The suitability of various materials for mRNA isolation and possible fractionation was examined by comparing the chromatographic behavior of synthetic poly(A) and rRNA on each support [215]. Oligo(dT)-cellulose appears to be the most suitable support of those investigated for RNA isolation and possible fractionation (Table 12).

In addition to the high specificity of oligo(dT)-cellulose toward poly(A) sequences, numerous nonspecific interactions were observed for oligo(dT)-cellulose [216]. First, nonpolyadenylated nucleic acids bind and are eluted under conditions used to elute poly(A) RNA. Second, "tight" nonspecific binding occurs, in which poly(A) RNA fails to elute under conditions that dissociate A·T bonds. Hydrolysis is required to remove tightly bound RNA. Oligo(dT)-cellulose has a low capacity for both these types of binding and can readily be used with a heterologous RNA (e.g., bacterial). Third, indirect nonspecific binding can also occur. rRNA aggregates with poly(A) RNA and thus can bind indirectly to oligo(dT)-cellulose. After these aggregates are disrupted by treatment with DMSO and heat, poly(A) mRNA free of rRNA can be isolated. Methodology that is effective in preventing nonspecific and indirect binding of nonpolyadenylated RNA species has been discussed.

Table 12 Selectivity Factors (Sf) for Various Poly(A) Binding Materials[a]

Material	Salt (M)	rRNA capacity (mg/g)	Poly(A) capacity (mg/g)	Sf
Sigmacell	0.3 KCl	0.02	0.113	5.6
Sigmacell	0.5 KCl	0.06	1.65	27
Sigmacell	0.5 NaCl	0.028	0.152	5.4
Oligo(dT)-cellulose	0.1 NaCl	<0.004	0.61	>150
Oligo(dT)-cellulose	0.5 NaCl	<0.004	2.04	>510
Poly(U)-cellulose	0.1 NaCl	<0.004	0.087	>22

[a]The RNA capacity of poly(U)-cellulose was estimated to be the same as oligo(dT)-cellulose, since CF 11 cellulose was used in both cases. The selectivity factor (Sf) is defined as the ratio of poly(A) capacity to rRNA capacity.
Source: Ref. 215.

A method for the isolation of mRNA from polysomes has been described [217]. Polysomes are dissolved in a solution containing 0.5 M NaCl and sodium dodecyl sulfate and applied to an oligo(dT)-cellulose column. RNA species containing poly(A) sequences are retained by the column, whereas ribosomal proteins and other RNA species are washed off. The column is then eluted with a buffer not containing NaCl. mRNA from HeLa cells and from duck reticulocytes has been fractionated in this way.

A method for eluting mRNA from agarose gels has been described [218]. Chick ovalbumin mRNA and globin mRNA were electrophoresed on 2% agarose-6.0 M urea cylindrical gels at pH 5.5. The position of the RNA after electrophoresis was determined in stained gels, and corresponding sections of unstained gels were cut out. These slices were pulverized and layered on top of an oligo(dT)-cellulose column to which diffusing RNA was bound. The recovery of poly(A) RNA varied between 30 and 50%; however, the biological activity of the eluted RNA was retained as measured by in vitro translation in the wheat germ system. Efficient recovery of functionally intact mRNA from agarose gels via transfer to an ion-exchange membrane has been described [446].

An expedient procedure for preparing mRNA from mammalian cells has been developed [219]. It is totally devoid of phenol-based organic solvents and utilizes the deproteinizing ability of the chaotropic agents lithium chloride and guanidinium chloride. Special considerations were given to preventing RNase action during the preparation. For this purpose 3 M LiCl/4 M urea/1 mg per milliliter of heparin/15 mM EDTA was chosen as the principal deproteinizing agent. RNase activity was completely suppressed in this mixture. The preparation method was applicable to both polysomal and total cytoplasmic RNA. Poly(A) mRNA was isolated using an oligo(dT)-cellulose column. The isolated mRNA preparations were analyzed for their intactness by sucrose gradient centrifugation and agarose gel electrophoresis in the presence of a denaturant, methyl mercuric hydroxide. The messenger activities were tested in cell-free translation systems.

A simple procedure is described which permits a one-step separation of total plant nucleic acids into DNA, rRNA, sRNA, oligonucleotides, two poly(A)-containing fractions, and residual polysaccharides by chromatography on columns of Sephadex G-75, Sepharose 4B, and poly(U)-Sepharose 4B, coupled serially [220].

Various techniques are commonly employed in the isolation of specific mRNAs and reviewed [221]. The development and refinement of these techniques has been of central importance in the ability to isolate individual mRNAs. However, each of these

Table 13 Isolation of mRNAs Coding for Various Proteins from Different Cell Sources Using Template Chromatography

Coded protein	Cell source	Method	References
?	Spleens or lymph nodes of guinea pig skin	Oligo(dT)-cellulose	257
?	Erythroid enriched bone marrow cells of the rabbit	Poly(U)-Sepharose	258, 259
?	Wing imaginal disks of *Pienis brasicae*	Poly(U)-Sepharose	260
Major structural protein of the adenovirus type 2 (ad2)	Ad2-infected HeLa cells	Oligo(dT)-cellulose	515
N polypeptide of vesicular stomatitis virus (VSV)	Indiana serotype of VSV	Oligo(dT)-cellulose	516
Rat preputial gland β-glucuronidase	Preputial glands of sexually mature female rats	Oligo(dT)-cellulose	517
Antifreeze peptides	Winter flounder livers	Oligo(dT)-cellulose	518, 519
Calf preprochymosin	Mucosal layer of the fourth stomach of a freshly slaughtered suckling calf	Oligo(dT)-cellulose	520
Neuron-specific enolase 14-3-2	Rat brain	Poly(U)-Sepharose	521
Precursor of ornithine transcarbamoylase; precursor of β subunit of propionyl-CoA carboxylase and cystationine β-synthetase	Rat liver	Oligo(dT)-cellulose	522
?	Mature and juvenile *Fasciola hepatica*	Oligo(dT)-cellulose	523

Precursors of proline-rich protein	Human submandibular gland	Oligo(dT)-cellulose	524
?	Developing endosperms from ears of field-grown wheat	Oligo(dT)-cellulose	525
α and β Subunits of mouse thyrotropin	Mouse thyrotropic pituitary tumors	Oligo(dT)-cellulose	549, 550
Pre-β subunit of mouse thyrotropin	Mouse thyrotropic tumor	Oligo(dT)-cellulose	526
Interleukin 2	Human peripheral blood leukocytes	Oligo(dT)-cellulose	527
Prorennin	Mucous layer of two abomasum of newborn calves	Poly(U)-Sepharose	528
Interferon	Human fibroblasts strains 17/1 and FS-4	Oligo(dT)-cellulose	529
	Poly(rI). poly(rC)-induced and non-induced FS 11 cells	Oligo(dT)-cellulose	558
Gastrin	Gastrin-producing tumor	Oligo(dT)-cellulose	530
Human histocompatibility antigen HLA-B	RPMI 4265	Oligo(dT)-cellulose	531
Variant surface glycoprotein	*Trypanosma brucei*	Oligo(dT)-cellulose	532
α-Tubulin	Neonatal rat brain	Oligo(dT)-cellulose	533
Photoreaction center and light-harvesting antenna polypeptides	*Rhodospirillum rubrum*	Oligo(dT)-cellulose	351
Bovine corticotropin and β-lipotropin precursor	Neurointermediate lobe of bovine pituitaries	Oligo(dT)-cellulose and poly(U)-Sepharose	535, 536

Table 13 (Continued)

Coded protein	Cell source	Method	References
Globin	Friend cells	Poly(I)-Sephadex	538
Human globin	Reticulocytes	Oligo(dT)-cellulose	559, 560
α_{2U}-Globulin	Mature male rats	Oligo(dT)-cellulose	539
Soybean leghemoglobin	Soybean nodules	Oligo(dT)-cellulose	557
Variant surface glycoproteins	*Trypanosoma brucei*	DNA-DBM paper	541
β Subunit of follicle-stimulating hormone	Pituitary glands of castrate male sheep	Oligo(dT)-cellulose	561
Chicken lysozyme	Oviduct	Oligo(dT)-cellulose	562
Trypanosome variable surface antigen	*Trypanosoma brucei*	Oligo(dT)-cellulose	566
Transferrin	Chick oviduct	Oligo(dT)-cellulose	575
Conalbumin	Chick oviduct	Oligo(dT)-cellulose	575
α-Fetoprotein	Male rats	Oligo(dT)-cellulose	576
Dopamine β-hydroxylase	Rat pheochromocytoma cell line	Oligo(dT)-cellulose	579
Apo VLDL-II (aproprotein in avian very low density lipoproteins)	Liver of 4-week-old cockerels	Oligo(dT)-cellulose	578

methods may require modification in order to be applicable to a particular system. Some examples of the isolation of specific mRNA are described in the following section and summarized in Table 13.

II. mRNA FROM VARIOUS ORGANISMS

A. Virus

Vaccinia is a large DNA virus which replicates exclusively in the cytoplasma of infected animal cells. Poly(A) sequences in mRNA has been observed [222]. The purification of adenovirus-specific mRNA has been described [223] and a simple method for selection of RNA·DNA hybrids has been developed [224]. Cytoplasmic RNA prepared from adenovirus type 2 (ad2)-infected HeLa cells or from an ad2-transformed rat cell line was hybridized in solution to the complementary strands of ad2 DNA. The hybridization mixture was subsequently fractionated by chromatography on a Sepharose 2B column. The intact probe DNA as well as the RNA·DNA hybrids are excluded from the gel matrix and elute with the void volume. Nonhybridized RNA, in contrast, is included into the gel matrix and elutes as a broad peak well separated from the excluded fractions. Fractions corresponding to the void volume were collected and the RNA·DNA hybrids were denatured in 90% formamide. The selected RNA was separated from the DNA by chromatography on poly(U)-Sepharose. Restriction endonuclease fragments of DNA of large enough size to exclude them from the agarose column were also used for hybridization. In these experiments hybridizations were carried out under conditions which would allow R-loop formation and the hybridized RNA was separated from unhybridized RNA by Sepharose chromatography. The validity of the method was demonstrated by programming an in vitro protein-synthesizing system with selected RNA.

Two major size classes of mammary tumor virus (MTV) RNA, 35S and 24S RNA, were isolated from MTV-infected rat (XC) cells and cultured mammary tumor cells by preparative hybridization of whole cell or poly(A) RNA to cloned MTV DNA covalently bound to chemically activated paper disks [225].

A method for purifying RNA by hybridization to complementary DNA immobilized on Sephadex has been described and illustrated by the purification of polyoma mRNA [226]. Polyoma DNA, or restriction fragments thereof, were bound to Sephadex and used to purify viral mRNA. Viral mRNA could also be purified by hybridization directly from total cytoplasmic infected cell RNA without prior selection of poly(A) RNA. The advantages and limitations of DNA attached to CNBr Sephadex as an matrix for RNA purification are discussed.

The isolation of mRNA from a C-type virus-positive cell (JLS-V6) has been described [227]. The isolation consisted of specific immunochemical precipitation of polyribosomes synthesizing viral proteins, extraction of the mRNA from the polyribosome/antigen/antibody complex, and separation of poly(A) RNA by chromatography on oligo(T)-cellulose. Three size classes of virus-specific poly(A) mRNA were identified.

Herpes simplex virus type 1 DNA covalently bound to cellulose was used to isolate and translate in vitro herpes simplex virus type 1 mRNAs [228,229]. Readily detectable, late mRNAs 2.7 and 1.9 kb in size encoding 69,000- and 58,000-dalton polypeptides, respectively, were isolated. A very minor late mRNA family composed of two colinear forms, one 2.6 kb and one 2.8 kb, was isolated and found to encode only an 85,000-dalton polypeptide. A major early mRNA, 1.8 kb in size encoding a 64,000-dalton polypeptide, was also isolated.

B. Yeast

Poly(A) sequences are found covalently linked to yeast mRNA molecules [230]. The poly(A) sequences are relatively homogeneous in size and have an average chain length of about 50 residues. They are located at the 3' end of the mRNA molecule. The data suggest that some mRNA molecules may not contain poly(A) sequences and that no poly(A) sequences are present in rRNA.

Yeast poly(A) mRNA was isolated from total cellular RNA by chromatography on poly(U)-cellulose [231]. The relative complexity of the isolated yeast mRNA was assessed by hybridization analysis with cDNA synthesized from the isolated mRNA with viral reverse transcriptase. Poly(A) yeast mRNA was translated in vitro in a wheat germ cell-free extracts, and the major polypeptides synthesized have the same molecular weight as the major proteins present in the cell. Four of these proteins were identified by coelectrophoresis and immune precipitation to pyruvate kinase, enolase, aldolase, and glyceraldehyde-3-phosphate dehydrogenase. These data demonstrate, in agreement with the hybridization results, that yeast contains major mRNA species and that some of the glycolytic enzyme mRNAs make up part of the major fraction. A procedure is outlined for the preparation of yeast mRNA which is essentially free of rRNA contamination and is further enriched in the major mRNAs present in the cell.

Radioactive mRNA was isolated from a *S. cerevisiae* strain containing the L double-stranded (Lds) RNA species [232]. This mRNA was hybridized to Lds RNA isolated from the same strain. Analysis of the hybrid formed shows it to be of a similar size to the Lds RNA. It is concluded from this result that the mRNA complementary to the -ve strand of the Lds RNA is polycistronic. When the total mRNA is fractionated on oligo(dT)-cellulose, RNA complementary to the -ve strand of Lds RNA is found only in the nonbinding fraction. When Lds RNA is fractionated on oligo(dT)-cellulose the majority is found in the nonbinding fraction. It is concluded that neither the Lds RNA genome nor the mRNA it encodes have poly(A) tails of significant length.

C. *Tetrahymena*

Exponentiall growing *Tetrahymena pyriformis* organisms were labeled with [^{3}H]uridine or [^{3}H]adenosine. The labeled RNA was extracted and isolated by chromatography on poly(U)-Sepharose and further analyzed by means of sucrose gradient centrifugation and RNase digestion [233]. Experimental evidence proved the existence of RNase-resistant poly(A) fragments in the RNA of *Tetrahymena* cells.

Experiments designed to compare products of translation of mRNAs of membrane-bound and free polysomes in *T. pyriformis* have been described [479]. The results show that a large number of the polypeptides produced in vitro are mainly coded by poly(A) mRNA of free polyribosomes, while a smaller number are specifically coded by poly(A) mRNA from membrane-bound polysomes.

D. *Dictyostelium discoideum*

mRNA from the cellular slime mold *Dictyostelium discoideum* contains a sequence of about 100 adenylic acid residues at the 3' end [455]. Most newly synthesized poly(A) RNA from this organism contains poly(A) tracts which shorten with age, and these new transcripts can be purified at least 10-fold, and perhaps as much as 60-fold, from other cellular mRNAs by thermal elution from poly(U)-Sepharose [234,456].

E. Plant

Polysomes of high molecular weight have been isolated from the leaves of light-grown pea seedlings previously labeled for 2 hr with either [^{3}H]adenosine or [^{3}H]uridine. These polysomes have been shown by both equilibrium and velocity centrifugation to be free of nonpolysomal RNA. Not more than 36% of the labeled mRNA associated with these polysomes binds to poly(U)-Sepharose and this mRNA is shown to contain poly(A) sequences which, by polyacrylamide gel electrophoresis, are shown to vary in size from larger than 5S-8S RNA to smaller than 4S RNA [235]. In the mRNA fraction that does not bind to poly(U)-Sepharose, negligible poly(A) sequences were detected. Cell-free protein synthesis studies illustrate that the two classes of mRNA code for the same major polypeptides.

A fraction of the RNA species isolated from *Lemma gibba* G-3 consists of molecules with attached sequences of poly(A) [236]. This poly(A)-containing fraction, separated from total RNA by adsorption onto oligo(dT)-cellulose, was shown to be mRNA by its ability to serve as template in a cell-free translation system derived from wheat germ. This method permitted the comparison of mRNA from plants grown under different light conditions. Such plants were shown to possess qualitative and quantitative differences in their mRNA complements.

The total poly(A) RNA from aerated carrot root disks was further fractionated into a C-rich mRNA fraction by oligo(dG)-cellulose chromatography [237]. C-rich mRNA was purified at least 10-fold by this procedure. Translation in double-label amino acid mixtures indicates that C-rich mRNA codes for proline-rich proteins which may be precursors to the hydroxyproline-rich glycoprotein synthesized in vitro by this tissue.

F. *Eimeria tenella*

RNA from oocytes of *Eimeria tenella* was purified into poly(A)$^-$ RNA and poly(A)$^+$ RNA fractions with chromatography on oligo-(dT)-cellulose [238]. In vitro translation assays showed that the level of translatable mRNA in the poly(A)$^-$ RNA fraction was negligible. Virtually all the translatable mRNA in *E. tenella* was polyadenylated: that is, it contained poly(A) tracts with more than 15 adenylate residues.

G. *Artemia salina*

Messenger activity of RNA species retained poly(A)$^+$ and nonretained poly(A)$^-$ by oligo(dT)-cellulose was assayed in cytoplasmic fractions of *Artemia salina* embryos using a wheat germ system [239]. In undeveloped embryos, mRNA activity was found in a fraction sedimenting at 15,000 × g and about half as much in the cytosol. The ribosome fraction, consisting mainly of 80S monomers, had no activity. After development for 15-16 hr, the total mRNA activity of the embryos increased three-fold. All three cell fractions contained mRNA activity in both poly(A)$^-$ and poly(A)$^+$ RNA. Over half of the total was in the polysome-containing fraction and about one-fifth in poly(A)$^+$ RNA. The most pronounced change with development was in the poly(A)$^+$ mRNA activity, which increased over sixfold. The patterns of polypeptides synthesized by translation of poly(A)$^+$ mRNA fractions from undeveloped and developing embryos were similar but not identical.

H. Mouse

Poly(A) mRNA from mouse kidney labeled in vivo exhibited a pattern of methylation distinct from that of rRNA and tRNA. After mice were given L-(methyl-^{3}H)methionine, 4% of the polyribosomal RNA label was bound to oligo(dT)-cellulose [240]; 20-24% of orotate- or adenine-labeled polyribosomal RNA eluted in the poly(A)$^+$ RNA fraction under similar conditions. [^{3}H]Methyl radioactivity was not incorporated into low-molecular-weight (5S-5.8S) rRNA. Based on an average molecular length of 2970 nucleotides, renal mRNA was estimated to obtain 8.6 methyl moieties per molecule. Although 83% of the [^{3}H]methyl radioactivity in rRNA was in the 2'-O-methylnucleotide fraction, no methylated dinucleotides were found in mRNA. In poly(A)$^+$ mRNA 60% of the [^{3}H]methyl label was in the mononucleotide fraction. Nominal poly(A)$^-$ mRNA and poly(A)$^+$ mRNA have complete nucleotide sequence homology, suggesting that kidney does not contain mRNAs that are not represented int he polyadenylated RNA fraction [484].

As a prerequisite to examining mRNA metabolism in compensatory renal hypertrophy, poly(A) RNA has been purified from normal mouse kidney polysomal RNA by selection on oligo(dT)-cellulose [241]. Poly(A) mRNA dissociated from polysomes by treatment with 10 mM EDTA and sedimented heterogeneously in dodecyl sulfate-containing sucrose density gradients with a mean sedimentation coefficient of 20S. Poly(A) derived from this RNA migrated at the rate of 6S-7S RNA in dodecyl sulfate-containing 10% polyacrylamide gels. Coelectrophoresis of poly(A) labeled for 90 min with poly(A) labeled for 24 hr indicated that the long-term labeled poly(A) migrated faster than pulse-labeled material.

Five or six percent (by mass) of AKR-2B mouse embryo cell polysomal RNA consists of mRNA sequences which may exist in polyadenylated form. In the steady state, however, only 30-40% of these molecules are retained by extensive passage over oligo-

(dT)-cellulose, the remainder being present in the form of poly(A)-deficient analogs [243]. These poly(A)-deficient analogs contain representatives of all poly(A) mRNA sequences in these cells. An analysis of the kinetics of hybridization of cDNA probes enriched for either abundant or rare poly(A) mRNA sequences suggests that the frequency distributions of poly(A) and poly(A)-deficient analogs are dissimilar, and that a relationship exists between the intracellular frequency of a given mRNA sequence and the number of poly(A)-deficient analogs of that sequence. High-frequency sequences appear to be enriched in the poly(A) fraction, while low-frequency sequences are associated predominantly with the poly(A)-deficient fraction; thus poly(A) may play a role in the regulation of mRNA frequency in the cytoplasm.

RNA of mouse liver was isolated and fractionated into poly(A)-containing and poly(A)-deficient classes by oligo(dT)-chromatography [244]. Approximately 99% of the poly(A) material bound to the oligo(dT); that which did not bind contained substantially shorter poly(A) chains. All RNA fractions retained the ability to initiate cell-free translation, with the poly(A)-deficient fraction containing half the total translational activity (i.e., mRNA). Two-dimensional polyacrylamide gel analysis of the cell-free translation products revealed three classes of mRNA.

I. Human Placenta

In order to understand better the regulation of human placental proteins, the activity of placental lactogen mRNA has been examined. Total RNA was extracted from normal-term placentas and purified by chromatography on oligo(dT)-cellulose [245,573]. The poly(A)-containing fraction stimulated amino acid incorporation 5- to 10-fold in wheat germ cell-free extracts, and immunoprecipitation of the translation products with antiserum directed against human placental lactogen (hPL) suggests that about 2% of the peptides contain hPL determinants. Analysis of the material precipitated with hPL antiserum by electrophoresis on sodium dodecyl sulfate-polyacrylamide gels revealed two major species. Tryptic peptide analysis of the large material and authentic hPL shows marked similarities in the primary structure of these two proteins. The slower-migrating peptide has an apparent molecular weight about 3000 larger than hPL and thus may represent a precursor molecule. Both cell-free products could be competed out of immunoprecipitates by a large excess of authentic hPL, confirming their immunologic similarities.

J. HeLa Cells

The appearance of newly formed mRNA in polyribosomes of HeLa cells is inhibited by over 85% by 3'-deoxyadenosine, probably due to the failure of normal attachment of poly(A) to heterogeneous nuclear RNA in the presence of this drug [246]. Results show that the labeled RNA that does reach polysomes in the presence of 3'-deoxyadenosine can be characterized as mRNA that contains smaller poly(A) segments than normal mRNA. The results of the present experiments suggest that all, or almost all, HeLa cell mRNA molecules (except for histone mRNA) are derived from nuclear RNA molecules that contain poly(A). The 3'-deoxyadenosine-sensitive step, presumably the addition of poly(A) to HnRNA, occurs in the cell nucleus a few minutes after the DNA-dependent synthesis of the RNA molecule destined to form mRNA.

The poly(A) sequences of about 150 nucleotides found among all size classes of the pulse-labeled RNA bound to HeLa polysomes, which is thought to be mRNA, provide a basis for the purification of such RNA by hybridization to oligo(dT)-cellulose [247,577]. Poly(A) sequences comprised 15% of the RNA from polysomes. This RNA sedimented in sucrose gradients as a heterogeneous population. A similar population of RNA molecules was also purified on oligo(dT)-cellulose from cytoplasmic components sedimenting more slowly than polyribosomes.

The poly(U) sequences of 30-40 nucleotides found in the HnRNA of HeLa cells are also present in the poly(A) mRNA of these cells [248]. The distribution of poly(U) sequences parallels that of poly(A) sequences in different size classes of cytoplasmic mRNA. Poly(U) sequences are also found in RNA molecules of the nucleus and cytoplasm which do not contain poly(A). Although poly(U) sequences are most abundant in nuclear RNA species lacking poly-(A), they are well represented in a class of RNA molecules in cytoplasm which lack poly(A), but which otherwise resemble the polysomal mRNA of HeLa cells.

The sedimentation properties of pulse-labeled and long-term labeled mRNA from highly purified HeLa cell free polysomes, selected for poly(A) content by two successive passages through oligo(dT)-cellulose columns, were analyzed under native and denatured conditions [249]. The sedimentation profile of the mRNA showed a broad distribution of components.

HeLa cell mRNA, isolated from mRNPs released from polysomes by EDTA to minimize contamination from nuclear RNA, was separated into four classes of molecules; those containing poly(A^+) as well as oligo(U^+) (4-13%); poly(A^+) oligo(U^-) (~52%); poly(A^-)

oligo(U^+) (~2%); and poly(A^-) oligo(U^-) (~33%) [250]. The oligo-(U) segments are 89% UMP and range from 20 to 50 nucleotides in length. The poly(A^+) oligo(U^+) mRNAs appear to contain the oligo(U) in a region with secondary structure since they did not bind to poly(A)-Sepharose without prior HCHO modification. HCHO modifies the exocylic amino groups of CMP, GMP, and AMP and prevents hydrogen bonding but reacts only slightly with UMP. After removal of the excess HCHO, the oligo(U) of the mRNA was free to bind to the poly(A)-Sepharose. The poly(A^+) oligo(U^+) RNAs appeared to contain one oligo(U) per molecule and could rebind to poly(A)-Sepharose with 31-75% efficiency. The poly(A^+) oligo(U^+) mRNA class was more heterogeneous in size and slightly larger than the other mRNA classes.

When cytoplasmic poly(A^+) RNA from HeLa cells was treated with RNase H and oligo(dT) to remove its 3'-poly(A) tail, increased binding to poly(A)-agarose was observed [251]. The bound material, which comprised 4-6% of the initial RNA, contained 65-80% of the oligo(U) sequences generated by RNase T_1 digestion. Oligo(U) isolated from the bound fraction was shown to be 83% U and to have a U/G ratio of 33. In contrast, oligo(U) from the unbound material was 77% U and had a U/G ratio of 13, suggesting that it is shorter and less U rich than the oligo(U) in the bound fraction. On sucrose gradients, oligo(U^+) RNA consistently sedimented with a larger S value than oligo(U^-) RNA. The oligo(U) content of oligo(U^+) RNA suggests one oligo(U) tract of 33 nucleotides per RNA molecule of 2000-3000 residues.

K. Ehrlich Ascites Cells

Polynucleotides containing AMP in excess of 90 mol % account for almost 1% of the total RNA synthesized by Ehrlich ascites cells growing either in the mouse or during short incubation in vitro. The affinity of these AMP-rich polynucleotides for oligo(dT)-cellulose has been used to isolate them from the total cellular RNA [252]. The most highly purified fractions are relatively large polynucleotides of the order of 8S-10S. They are located primarily in the cell nucleus, for almost 90% of the polyadenylate of whole cells was recovered in nuclei from an aqueous homogenate of these cells.

L. KB Cells

mRNA from polysomes of KB cells was isolated by chromatography on poly(U)-Sepharose [79]. The presence of poly(A) RNA in the nonpolysomal fractions of the cytoplasm was demonstrated. This putative mRNA was shown to constitute about 30% of the total cytoplasmic poly(A) RNA. Also by using the poly(U)-Sepharose it was demonstrated that 0.04 μg per milliliter of actinomycin D supresses the appearance of mRNA on polysomes up to 50%. This low concentration of the drug was previously thought to effect only rRNA synthesis.

M. Myeloma Cells

Poly(A) RNA has been isolated from the nucleus and cytoplasm of mouse plasmacytoma cells (5563) [253]. Injection of this RNA into *Xenopus* eggs and oocytes has demonstrated that both the nuclear and cytoplasmic RNA contains messages for the immunoglobin heavy and light chains.

Crude preparations of biologically active mRNA, which code for a myeloma (MPC-11) light chain, were isolated by two successive sucrose gradient centrifugations of RNA extracted from membrane-bound ribosomes. mRNA thus obtained was separated into a poly(A)-rich and a poly(A)-poor fraction by oligo(dT)-cellulose chromatography [254]. Both these fractions were able to direct the synthesis of light chains in reconstituted cell-free systems. The identity of the products in vitro was confirmed by comparing their migration with that of light chains produced in vivo upon electrophoresis in sodium dodecyl sulfate/polyacrylamide gels, and from the profiles of tryptic peptides obtained by chromatography on Aminex A-5 ion-exchange columns. Template activity of the poly(A)-rich light-chain mRNA fraction showed very little variation during the cell cycle. The activity of the poly(A)-poor fraction, on the other hand, was maximal during the early S phase. It is concluded that maximal synthesis of immunoglobulins observed in vivo during the late G_1 phase of the cell cycle is achieved by translational control mechanisms.

N. Novikoff Hepatoma

Excess cDNA from regenerating liver or Novikoff hepatoma, covalently linked to cellulose, was used to adsorb the complementary mRNA sequences from Novikoff hepatoma or regenerating liver [255]. As shown by two-dimensional gel electrophoresis, the translated products of the bound mRNA fractions contained proteins common to both tissues.

O. Friend Leukemia Cells

Utilizing the technique of poly(A)-Sepharose chromatography, it is possible to isolate a novel class of RNA molecules from polysomes of *Friend* leukemia cells [256]. These RNA species display mRNA-like behavior. They are released from polysomes on treatment with EDTA and are able to direct polypeptide synthesis in a cell-free protein synthesizing system. They appear to be distinct from the poly(A) mRNAs, as judged by their lack of a 3'-terminal poly(A) tract, by their different size distribution, by their unusual base composition, by the presence of a possible "uridylate-rich" region toward their 3' end, by their low sequence homology to poly(A) mRNAs, and by the difference in at least some of their translation products.

When total cytoplasmic RNA from mouse Friend cells is fractionated using oligo(dT)-cellulose or poly(U)-Sepharose chromatography, approximately 20% of the mRNA activity remains in the unbound fraction, even though this contains 0.5% of the poly(A) [242]. This RNA, operationally defined as poly(A)$^-$, is found almost entirely in polysome structures in vivo. Its major translation products are the histones and actin. With the exception of the mRNAs coding for histones, poly(A)$^-$ mRNA encodes similar proteins to poly(A)$^+$ mRNA, although in very different abundances.

III. TURNOVER OF mRNAs

Adenovirus-specific nuclear and polysomal RNA, both early and late in the infectious cycle, contain a covalently linked region of poly(A) 150-250 nucleotides long [261]. As revealed by hybridization experiments, the poly(A) is not transcribed from adenovirus DNA. Furthermore, an adenosine analog, cordycepin, blocks the synthesis of poly(A) and also inhibits the accumulation of adenovirus mRNA on polysomes.

The changes occurring in the pattern of translatable mRNA species in germ cells during spermatogenesis of mice has been investigated [511]. Highly homogeneous cell populations of spermatocytes or spermatids were purified. Poly(A)$^+$ mRNA was isolated from each cell population by oligo(dT)-cellulose chromatography. A comparison of meiotic and postmeiotic mRNAs was made by two-demensional gel analyses of their in vitro synthesized translocation products. Among spots identified in the fluorograms of two-dimensional gels, a number of qualitatively new proteins appeared after meiosis. The results suggest that some poly(A)$^+$ mRNAs are transcribed postmeiotically in haploid germ cells.

The levels of histone mRNA are rapidly reduced after treatment of cultured cells with hydroxyurea or cytosine arabinonucleoside [404]. The histone mRNA for the replicative histone variants is destroyed rapidly, with a half-life of 10-15 min. The levels of mRNA coding for the replacement histone variant H3.3 were unchanged after treatment with DNA synthesis inhibitors. In addition to the rapid destruction of histone mRNA, there was a reduction to one-fifth in the rate of transcription of the histone genes. Lymphoma cells (S49) arrested in G1 by cyclic AMP produce and contain significant levels of histone mRNA. Hydroxyurea reduces the rate of transcription and the levels of histone mRNA in the G1-arrested cells.

Poly(A) RNA was isolated from fission yeast (*Schizosaccharomyces pombe*) total RNA using oligo(dT)-cellulose, and was studied as a model for mRNA [262]. The half-life of poly(A) RNA, measured by two independent methods, is about one-third of the cell doubling time, and is much longer than previous estimates by indirect methods of yeast mRNA half-life. Both experimental methods provided evidence for the existence of two populations of poly(A) RNA. The larger population has a half-life of 40-50 min; a much smaller population is probably turning over more rapidly. After inhibition of RNA synthesis by 8-hydroxyquinoline, the rate of total protein synthesis declined much more rapidly than the poly(A) RNA content of the cells. However, 60 min after inhibition of RNA synthesis there was a small rise in the rate of protein synthesis.

The effect of aging on the properties of polysomal poly(A) mRNA of *P. polycephalum* has been investigated [263]. Using poly(U)-Sepharose chromatography it was shown that shortening of the poly(A) tract occurred as the age of the mRNA population increased. A reduction in the efficiency of utilization of mRNA for translation as the age of the mRNA population increased was demonstrated by measuring the proportion of poly(A) mRNA present in the polysomal fraction compared with postpolysomal material.

Polysomes and their associated RNAs were extracted from aging soybean cotyledons [264]. Density gradient analysis showed that the ribosomes of older cotyledons were present largely as monosomes and light polysomes, while younger cotyledons possessed a high proportion of heavy polysomes. Increased polysomal and mRNA activities were observed in 3- to 5-day-old and again in 10-day-old cotyledons. The presence of poly(A) mRNA was determined by oligo(dT)-cellulose chromatography (Fig. 12) and confirmed by hybridization with [^{3}H]poly(U).

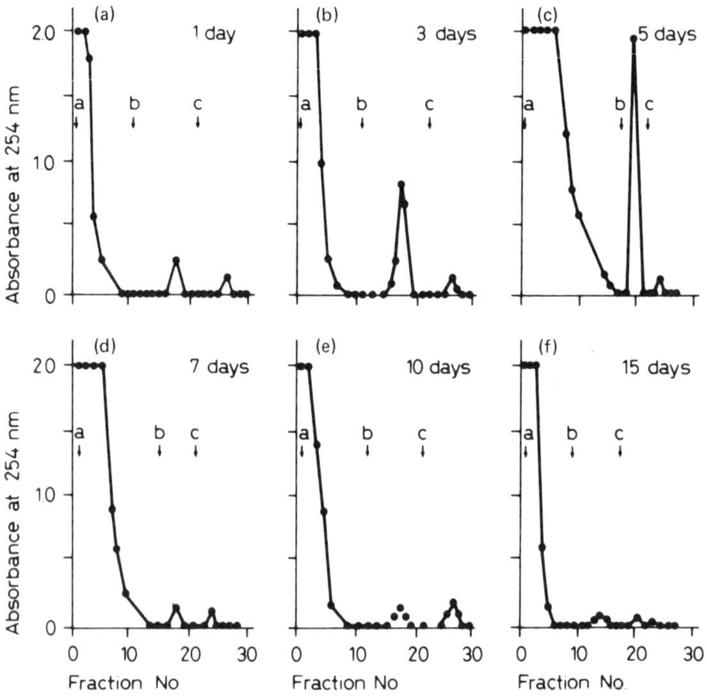

Figure 12 Fractionation of RNA from polysome pellets of soybean cotyledons aged 1, 3, 5, 7, 10, and 15 days by oligo(dT)-cellulose chromatography. RNA was phenol extracted from the polysome pellets and 160 A_{254} units were loaded onto on oligo(dT)-cellulose column. The buffers were changed as indicated by the arrows to elute the various fractions. a = 0.5 M KCl-Tris buffer; b = 0.01 M KCl-Tris buffer; c = 0.1 M Tris-HCl, pH 7.5. (From Ref. 264. Copyright 1982 S. Karger AG, Basel.)

IV. mRNA OF VARIOUS PROTEINS

A. Globin

Mammalian globin mRNA was the first eukaryotic mRNA to be isolated and studied extensively. A convenient technique (Table 13) for the partial purification of large quantities of functional poly(A) mRNA is described [265]. The method depends on annealing poly(A)

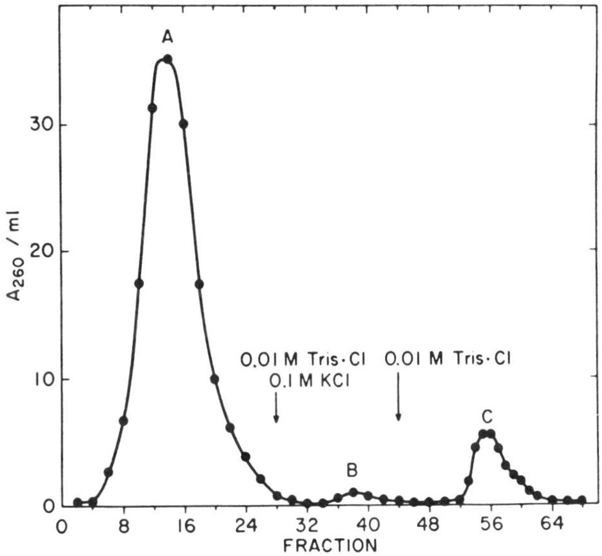

Figure 13 Oligo(dT)-cellulose chromatography of reticulocyte polysomal RNA. Peak A, material that is not retained by the column; peak B, material eluted with the indicated buffer of intermediate ionic strength; peak C, material eluted with the indicated buffer of low ionic strength. (From Ref. 265.)

mRNA to oligo(dT)-cellulose column (Fig. 13) and its elution with buffers of low ionic strength. Biologically active rabbit globin mRNA has been purified by this procedure and assayed for its ability to direct the synthesis of rabbit globin in a cell-free extract of ascites tumor. Reticulocyte polysomes could be dissolved in sodium dodecyl sulfate and applied directly to an oligo(dT)-cellulose column without prior extraction of protein. Examination of the eluted mRNA fraction showed that 9S globin mRNA also occurred in specific larger aggregates with 18S and 28S rRNA [266, 267]. These translationally active larger forms could be readily dissociated by heating at 65°C for 10 min to yield 9S globin mRNA and rRNA.

Rabbit globin mRNA was purified using oligo(dT)-cellulose and sucrose gradient centrifugation [268]. Both α- and β-globulin mRNA molecules behave heterogeneously with respect to their elution properties during chromatography on oligo(dT)-cellulose.

Those fractions eluted at the lowest ionic strength are most active in directing cell-free globin biosynthesis. Since >98% of the globin mRNA sequences are bound to oligo(dT)-cellulose, it was suggested that most polysomal globin mRNAs contain a poly(A) region and that this region is not of uniform length nor preponderately associated with either the α- or β-globin mRNAs.

Rabbit globin mRNA species containing poly(A) segments of different lengths were prepared by partial phosphorolysis of mRNA with *E. coli* polynucleotide phosphorylase [269]. It was found that the functional stability of the mRNA species containing 32 or more adenylate residues after injection into *Xenopus laevis* oocytes equaled that of the native globin mRNA. On the other hand, the functional stability of mRNA containing an average number of 21 adenylate residues was about 30% of the native mRNA, while that of mRNA containing 16 adenylate residues was as low as poly(A)-free globin mRNA.

Using polyacrylamide gel elution-electrophoresis in aqueous medium, highly purified rabbit globin mRNA can be fractionated into several populations of molecules differing by their mean poly-(A) content [270]. Both α- and β-globin mRNA are heterogeneous with respect to their electrophoretic mobilities. With the conditions used, no separation of α- and β-globin mRNA occurs during electrophoresis. From the specific radioactivity distribution in the different mRNA fractions one can conclude that the poly(A) sequence at the 3' end of globin mRNA molecules becomes shorter with aging. Furthermore, using two different methods of mRNA fractionation (polyacrylamide gel elution-electrophoresis and elution of poly(U)-Sepharose-bound mRNA at increasing temperatures), it is shown that old mRNA molecules differ from relatively young messages in their ability to direct cell-free globin synthesis. Modifications reducing template activity in vitro thus seem to take place during globin mRNA aging.

Melted double-stranded RNA isolated from rabbit bone narrow pre-mRNA was hybridized with excess of globin mRNA [271]. Seven to nine percent of double-stranded sequences became RNase-stable and approximately 30% of the sequences could be bound to poly(U)-Sepharose through poly(A) of mRNA. The size of RNase-stable hybrid is approximately 30 nucleotides, that is, one-fourth of the length of one strand of the double-stranded RNA.

Human globin mRNA prepared from erythroid cells of patients with sickle-cell anemia has been translated in *X. laevis* oocytes [272]. Addition of hemin to the injected mRNA causes total globin synthesis to increase and the ratio of α- to β^S-globin synthesis

(α/β^S ratio) to approach unity. To determine the effect of the length of the poly(A) on human globin mRNA stability, 10S globin mRNA was fractionated into poly(A)$^+$ and poly(A)$^-$ fractions by oligo(dT)-cellulose column chromatography. When oocytes are injected with each of these fractions, translation of the poly(A)$^+$ globin mRNA is sustained for a longer period than that of the poly(A)$^-$ mRNA. Regardless of the mRNA fraction injected, the α/β^S ratio of the synthesized globin decreases as the injected oocytes are incubated for longer periods. The results indicate that in frog oocytes poly(A)$^+$ mRNA has greater translational stability than poly(A)$^-$ mRNA, and β-mRNA has greater stability than α-mRNA with comparable poly(A) content.

The 5-terminal structures of murine α- and β-globin mRNA were determined after incubating cells of the erythropoietic spleen of mice with methyl-[^{3}H]methionine [273]. Globin mRNA was obtained from total cellular RNA by oligo(dT)-cellulose chromatography followed by elution of mRNA from formamide gels after electrophoresis. The globin mRNA was then hydrolyzed with KOH or digested with a combination of RNase T_2 and bacterial alkaline phosphatase, and 5'-terminal structures were isolated by DEAE-cellulose chromatography. The methylated nucleotides of these 5' structures were determined following digestion with specific RNase and bacterial alkaline phosphatase. Analyses of mRNA fractions enriched for either α- or β-mRNA gave similar results.

Studies were performed with duck immature red blood cells to identify and characterize the globin mRNA sequences in nuclear RNA [274]. Annealing of [^{3}H]globin cDNA to unlabeled nuclear RNA has identified three distinct size classes of nuclear RNA molecules containing globin mRNA sequences. The largest size class contained 1-2% of total nuclear globin mRNA sequences and sedimented through 85% formamide-sucrose gradients at the same rate as 28S rRNA. Chromatography on oligo(dT)-cellulose indicated that most of these molecules are not polyadenylated. The bulk of nuclear globin mRNA sequences (70%) was contained in poly(A) RNA molecules which sedimented at 16.5S. The remainder of the nuclear globin mRNA sequences (~30%) were detected in molecules sedimenting at 10S (the position of cytoplasmic globin mRNA). To determine whether a precursor-product relationship exists between these nuclear molecules and cytoplasmic globin mRNA, pulse-label and chase experiments were performed. Labeled globin mRNA sequences were assayed by annealing to globin cDNA-cellulose. Labeled 28S nuclear globin RNA sequences could not be detected, perhaps for technical reasons. The 16.5S nuclear globin RNA was labeled and chased into cytoplasmic globin mRNA sequences. The half-life of 16.5S nuclear globin RNA was estimated to be <30 min. These results demonstrate that in duck immature red blood cells, globin mRNA is transcribed as a larger precursor. Furthermore, size characterization of this precursor during pulse-label and chase periods suggests that it is processed within the nucleus to 10S globin RNA.

Pigeon mRNA coding for globin synthesis contains a poly(U) sequence of 30-40 nucleotides, as is shown by chromatography on poly(A)-Sepharose, chromatography on poly(U)-Sepharose of DNA synthesized on the mRNA by reverse transcription, and hybridization of [^{3}H]uridine-labeled mRNA with poly(A) [275]. The poly(U) sequence is in the noncoding part of the mRNA at the 3' end at a distance of no more than 100 nucleotides from the poly(A) sequence. The presence of the complementary sequences may lead to cyclization of the 3' region, with resultant protection against 3'-exonuclease.

Erythroblasts, obtained from the bone marrow of anemic chickens, were pulse labeled with [^{3}H]uridine, and total cellular RNA was isolated [276]. [^{3}H]pulse-labeled erythroblast RNA contains a 14S precursor to cytoplasmic 10S globin mRNA, and the precursor has a short half-life of about 3 min at 39°C. Properties of this RNA species satisfy the criteria for a functional higher-molecular-weight precursor to polysomal globin mRNA. It is higher in molecular weight than polysomal globin mRNA. It is newly synthesized; the precursor is processed to 10S globin mRNA. In addition, the precursor binds to oligo(dT)-cellulose, and so is polyadenylated. Newly synthesized globin RNA sequences were detected by cDNA hybridizations to different size-class fractions of erythroblast RNA, separated by sucrose gradient centrifugation. Globin RNA hybridization demonstrated that steady-state globin mRNA within this gradient was centered around 10S. Hence the 14S globin RNA species is not a product of 10S globin RNA aggregation. RNA hybridizations were also used to probe for globin RNA sequences in the size range 20-50S. In contrast to other reports, no 20-50S globin RNA sequences were detected.

A three-step procedure is described for the purification of globin mRNAs and their precursors from nucleated mouse erythroid cells [458]. The steps are chromatography on oligo(dT)-cellulose, electrophoresis in a 4.5% polyacrylamide gel containing 98% formamide, and hybridization to globin cDNA-cellulose.

The coding sequence of globin mRNA has been located at or very close to the 3' end of its poly(A)-containing 16S precursor [485]. The 16S RNA was annealed to globin cDNA and the hybrid digested with ribonuclease H. The undigested fragment did not bind to oligo(dT)-cellulose and its size was that expected for the intact 5' portion of the precursor.

The complete chromosomal mouse β^{maj}-globin gene, including its intervening and flanking sequences, has been cloned in the monkey virus SV40. The mouse gene is transcribed, processed, and translated in infected monkey kidney cells to yield mouse β^{maj}-globin [278].

B. Ceruloplasmin

Partially purified ceruloplasmin mRNA was isolated using indirect immunoprecipitation of rat liver polysomes and poly(U)-Sepharose chromatography of polysomal RNA [279]. This RNA programmed the synthesis of ceruloplasmin polypeptides in a cell-free system from mitochondria. Immunochemical analysis of the translation products revealed a 40-fold enrichment of the ceruloplasmin mRNA activity.

C. Uteroglobin

Uteroglobin, an hormonally induced protein composed of two similar subunits, represents approximately 50% of the proteins synthesized and secreted into the uterine lumen of rabbits treated sequentially with estradiol and progesterone. The endometrium of these animals was used as a source for the isolation of the mRNA for uteroglobin [280]. Poly(A) RNA, extracted from purified polysomes with phenol chloroform and isolated on oligo(dT)-cellulose columns, contains one-fourth of the total protein coding activity of the endometrium. Between 20 and 25% of the polypeptides synthesized by this RNA in cell-free systems derived from Krebs II ascites cells or wheat germ react with a monospecific antiserum prepared in guinea pigs against uteroglobin. The material bound to the antibody was identified as a precursor of uteroglobin.

D. Mouse Immunoglobulin

The mRNA coding for a mouse immunoglobulin L chain was prepared from MOPC-321 myeloma polysomes specifically precipitated with antibodies directed against L chains, followed by chemical purification on oligo(dT)-cellulose [281,556]. Biological purity was calculated to be >95%. Chemical purity (95%) was determined from the amount of rRNA in the mRNA preparation by scanning of appropriate gels. In a cell-free system, the purified mRNA directed the synthesis of two precursors heavier than L chain by about 1300 and 4700 daltons. The main fraction of L-chain mRNA was composed of two species of about 420,000 and 450,000 daltons. These molecules are much larger than that required to code for a mature L chain (calculated as ~250,000). The highly purified (>95%) mRNA coding for immunoglobulin L-chain yields on acrylamide gels a discrete 15.5S band, and a "shoulder" was isolated and found to be fragmented mRNA [282]. Partial cleavage of the mRNA probably occurs during mechanical disruption of the myeloma cells. Fragments with +intact 3' ends are selected due to binding via the poly(A) moiety to oligo(dT)-cellulose; that is, the fragments should be deficient at the 5' end, where mRNA translation is initiated. In agreement, the fragmented mRNA is essentially untranslatable in a cell-free system. The size of the L-chain mRNA is rather uncertain. A value of 15.5S is obtained from migration in acrylamide gels made in water or formamide, a value of 12S is obtained from sucrose gradient centrifugation.

The 500-fold purification of an mRNA encoding an immunoglobulin λ light chain derived from the mouse myeloma tumor, RPC-20, has been described [283]. Purification involves the isolation of membrane-bound polysomes, oligo(dT)-cellulose chromatography, and sucrose gradient centrifugation under conditions favoring denaturation of polynucleotide complexes. The mRNA purified in this way directs the cell-free synthesis of a polypeptide which is five or six amino acids longer than the mature form of RPC-20 light chain.

A mutant cell line (IF2) derived from the mouse myeloma MOPC 21 has been used for the isolation and sequence analysis of H-chain mRNA [284]. The IF2 cells synthesize an H chain of reduced size in which the C_H1 homology region is missing. Sizing of the IF2 H-chain mRNA and wild-type H-chain mRNA revealed that the deletion is expressed at the mRNA level. The mutant H-chain mRNA sedimented at 16S, enabling effective resolution from 18S rRNA. In experiments using IF2 cells labeled with [^{32}P]phosphate, the 16S mRNA was purified by oligo(T)-cellulose chromatography. Polyacrylamide gel analysis of the poly(A)-containing fraction showed the presence of a single radioactive band. Comparison of the mobility of this band relative to markers of known molecular weight revealed that the molecule contained about 1600 nucleotides. By using the established amino acid sequence of the IF2 H chain and a knowledge of the genetic code, 14 oligonucleotides were assigned within the constant region and 4 within the variable region of the IF2 H chain. These sequence data account for 19.5% of the coding region. Several other oligonucleotides, which could not be assigned within the coding region but which occurred in approximate molar yield, have also been partially characterized. These oligonucleotides are presumably derived from the untranslated regions of the mRNA.

A procedure is described for the isolation of highly purified H-chain immunoglobulin mRNAs from a variety of mouse plasmacytomas (IgA, IgG, and IgM producers) [534,540]. The use of fresh tissue and the rapid isolation and direct extraction of membrane-bound polyribosomes were found to be essential in obtaining large quantities of undegraded H-chain mRNAs. The individual mRNAs were purified by two cycles of oligo(dT)-cellulose chromatography, sodium dodecyl sulfate-sucrose gradient centrifugation, and electrophoresis on 98% formamide containing polyacrylamide gels. When added to a cell-free protein-synthesizing system from wheat germ, the MPC-11 γ_{2b} and H2020 α heavy-chain mRNAs efficiently directed the synthesis of a predominant product of 55,000, while the synthesis of a 70,000-dalton protein in addition to other lower-molecular-weight polypeptides were observed with MOPC 3741 μ mRNA. The purities of the α, γ_{2b}, and μ mRNAs were estimated to be 60-80%, 50-70%, and 50-83%, respectively, on the basis of their hybridization rates with cDNA probes in comparison to mRNA standards of known complexity. Heavy-chain mRNAs of the same class isolated from different mouse strains (Balb/C or NZB) display no detectable sequence differences in cross-hybridization experiments, even though the cDNA-mRNA hybrids are submitted to stringent S_1 nuclease digestion.

A procedure has been described for the large-scale purification of L- and H-chain mRNAs from plasmacytomas produced in mice [285]. L- and H-chain mRNAs were purified by oligo(dT)-cellulose chromatography and/or sucrose gradient centrifugation. γ2a and α H-chain mRNAs sedimented as major components at 15.5S and 16.5S, respectively, when L-chain mRNAs sedimented as 12S species. H-chain mRNAs isolated by continuous elution during preparative gel electrophoresis were completely separated from both L-chain mRNA and residual 18S rRNA; and migrated as single components of 1900 $\pm$ 50 nucleotides on analytical denaturing gels. The partially purified H-chain mRNAs were translated into major components of molecular weight 56,000 (γ2a) and 60,000 (α) in an mRNA-dependent rabbit reticulocyte lysate, whereas L-chain mRNAs yielded polypeptides of molecular weight 25,000 (λ) and 27,000 (κ). Up to 95% of the translation products directed by the purified mRNAs were immunoprecipitated using specific antisera. The purity of L- and H-chain mRNAs was assessed by hybridization of corresponding cDNAs with excess recombinant plasmid DNA. The results indicated a minimum purity of 47% (γ2a), 62% (α) for H-chain mRNAs, and 60% (κ) for L-chain mRNAs.

The cDNA to a mouse immunoglobulin μ H-chain mRNA has been cloned into the PstI site of the plasmid vector pBR322 [286]. A hybrid plasmid pμ 183.5 containing a 1850-base pair (bp) insert has been selected by differential screening. The nucleotide sequence of the insert encodes the four constant domains, the terminal piece, and the 3' untranslated region.

The cDNAs to mouse immunoglobulin α H-chain mRNAs have been cloned into the PstI site of the plasmid vector, pBR322 [287]. Recombinant plasmids have been identified by hybrid-arrested translation and purification of α H-chain mRNA on DNA-DBM filters. The nucleotide sequence of the inserts encodes the constant and 3' untranslated regions of the α H-chain mRNA. The CH3 domains of human and mouse α chains are highly homologous, including a 36-amino acid fragment not reported in the protein sequence. As in the case of the μ secreted H-chain, the α H-chain contains a carboxy-terminal piece of 20 amino acids.

Three cDNA clones for β_2-microglobulin, the small subunit of the major histocompatibility antigens, have been isolated [288]. β_2-Microglobulin makes up less than 0.1% of mouse liver protein, and its mRNA is approximately 0.03% of liver poly(A) mRNA. The cDNA clones were identified by screening 1400 cDNA clones made from 9S-10S mouse liver poly(A) mRNA. The procedure for screening the cDNA clones involved binding pooled plasmid DNA to nitrocellulose filters and testing the ability of each filter to select β_2-microglobulin mRNA. The filter-selected mRNAs were assayed for their ability to direct the synthesis of β_2-microglobulin in translation reactions in vitro. The isolated clones were shown by nucleotide sequence analysis to encode β_2-microglobulin.

Microsomal RNA from SL2 cells was fractionated on oligo(dT)-cellulose into three fractions: I was very similar to the nonfractionated RNA; II contained 28S and 18S rRNA; and III, [poly(A)-RNA] of 4S-50S, constituted 1.0-2.1% of the total RNA pool and was still contaminated with rRNAs [512]. The last fraction was synthesized at the highest rate. Anti-H-2^d serum precipitated cell-free translation products of 17S (heavy chain) and 10S (β_2-microglobulin) mRNA fractions obtained by centrifugation on 10-30% linear sucrose gradient. β_2-Microglobulin synthesis was fivefold higher as compared with the H-chain synthesis. With liver mRNA, the patterns of translation products were identical, but β_2-microglobulin synthesis was lower by a half than H-chain synthesis. H-2^d H-chain messenger constituted 0.3-0.4% of the enriched fraction and was purified 15- to 20-fold.

E. Myosin

Myosin mRNPs were prepared from 12-day embryonic chick muscle [289]. The ribosome mRNP fraction was phenol extracted, and then precipitated in ethanol, the poly(A) mRNA being isolated from total RNA by its ability to bind to oligo(dT)-cellulose. The poly(A) RNA, eluting from the column at low salt, has a major peak with the characteristic sedimentation value of myosin mRNA. From 40 to 45 g of tissue about 15-20 μg of this 26S mRNA can be obtained. However, subsequent examination by RNA·cDNA hybridization kinetics suggested that the 26S fraction was composed of two different mRNA sequences [290].

F. Collagen

Procollagen mRNA was isolated from the calvaria of 15-day-old chick embryos by chromatographing total RNA twice over oligo-(dT)-cellulose and then fractionating the twice-bound RNA on sucrose gradients [291,551,552]. When analyzed on 99% formamide gels, the 27S-30S fraction had three sharp fluorescent bands. In wheat germ cell-free extracts, the 27S-30S fraction directed the synthesis of two prominent collagenase-sensitive polypeptides with mobilities corresponding to the calvaria pro-α-chain markers. Twelve percent of this [^{3}H]proline-labeled wheat germ product could be hydroxylated with prolyl hydroxylase. Further purification was carried out by means of agarose adsorption chromatography. In a buffer containing a relatively high salt concentration, the guanine-rich procollagen mRNA preferentially bound to the agarose, with subsequent elution carried out at a low ionic strength [292,542].

The presumptive mRNAs for type I procollagen were isolated from chick embryo calvaria at various stages of development [293]. Poly(A) RNA fractions from denaturing sucrose gradients directed protein synthesis in a cell-free system derived from wheat germ. Procollagen mRNA activity was detected in a region of about 26S. Approximately 80% of the labeled proline incorporated into cell-free product was susceptible to digestion by purified bacterial collagenase. The synthesis of procollagen mRNAs was followed during development. The results suggest that procollagen mRNA is a relatively stable species with a prolonged half-life compared to the majority of mRNAs in this tissue.

RNA was extracted from fetal calf skin using phenol or guanidine hydrochloride [294]. Poly(A) RNA was separated by oligo(dT)-cellulose chromatography and was further fractionated by sucrose density gradient centrifugation. When translated in an optimized wheat germ extract cell-free system, unfractionated guanidine hydrochloride-extracted poly(A) RNA directed the synthesis of two collagenase-sensitive protein bands, while phenol extracted poly(A) RNA with a sedimentation coefficient greater than 25S was the only fraction to direct the same synthesis. These proteins were identified with procollagen α1(I) and procollagen α2. Inhibition of translation by phenol-extracted poly(A) RNA with a sedimentation coefficient less than 25S was also observed. Guanidine-hydrochloride-extracted poly(A) RNA from fetal calf skin directed the synthesis of three distinct collagenase-sensitive proteins in the micrococcal nuclease digested rabbit reticulocyte cell-free system: these seemed to correspond to procollagen α1(I), procollagen α2, and procollagen α1(III).

Polysomes were isolated from both lung and dermis of neonatal rats and the poly(A) RNA therein was isolated by oligo(dT)-cellulose chromatography [295]. The RNA fractions were then translated in the nuclease-treated reticulocyte lysate in the presence of radioactive proline and directed the synthesis of both pro α_1 and pro α_2 chains. The poly(A) RNA isolated from both lung and dermis polysomes of rats treated with triamcinolone diacetate synthesized significantly less collagen peptide as determined by collagenase digestion as did the RNA isolated from polysomes of nontreated animals. Noncollagen protein synthesis was decreased less than was collagen synthesis. Glucocorticoid treatment did not affect the ability of either polysomes in wheat germ extract or polysomal poly(A) RNA in nuclease-treated reticulocyte lysate to synthesize prolyl hydroxylase as determined by immunoprecipitation. These data indicate that the glucocorticoid-mediated inhibition of collagen polypeptide synthesis does not result from a nonspecific effect on total cellular protein synthesis of normal fibroblasts but from a selective reduction of polysome-associated mRNA. Furthermore, these data provide a molecular basis for selective inhibitory effect of synthetic anti-inflammatory steroids on collagen synthesis.

G. Crystallin

The mRNAs for lens crystallins, which are the major structural proteins of the lens, have been isolated. The 14S mRNA, coding for the A_2 chain of α-crystallin, was isolated from calf lens polyribosomes by zonal centrifugation and further purified by oligo-(dT)-cellulose chromatography [296]. The purified mRNA was used as a template to synthesize cDNA by reverse transcriptase. The DNA product appeared to be smaller in size than the RNA template but a faithful transcript of base sequences in the 14S α-crystallin mRNA.

Chromatography of lens polyribosomal RNA on oligo(dT)-cellulose yields three fractions [297]. As a rule the second fraction has been neglected in other studies reported. This fraction, in particular, contains the messengers from the noncrystallin lens proteins. From both calf lens and rat lens material, 10S and 14S RNA fractions were isolated that directed the cell-free synthesis of β-crystallin and α-crystallin, respectively [298,299].

H. Fibroin

Knowing the amino acid composition and sequence of the major peptides of silk fibroin, it may be inferred that there is a very high degree of reiteration (~2200) of the basic hexanucleotide -GCUGGU- within fibroin mRNA as well as within the gene. The synthesis of the hexanucleotide d(A-C-C-A-G-C), complementary to the two major triplets of fibroin mRNA, has been described [300]. In view of the ability of a prototype column, oligo(dC)-cellulose, to isolate fibroin mRNA, prospects appear excellent for the d(A-C-C-A-G-C)-cellulose column isolation of fibroin mRNA. Hydrophobic chromatography also permits the isolation of fibroin mRNA [301].

I. Elastin

mRNA was isolated from thoracic aortas of 16-day chick embryos by digestion of the tissue at 40°C with proteinase K in 1% sodium dodecyl sulfate followed by chromatography on oligo(dT)-cellulose columns [302]. Blunt-ended heteroduplex molecules consisting of one strand of mRNA and one of cDNA were formed using reverse transcriptase and S1 nuclease. The duplexes were tailed with dCTP and hybridized to the plasmid pBR322, which had been restricted with PstI and tailed with dGTP. The recombinant plasmids were used to transform *E. coli* C600. Ampicillin-sensitive and tetracycline-resistant colonies were screened by in situ hybridization with partially purified elastin mRNA that had been terminally labeled with [^{32}P]. Colonies that hybridized with the probe were selected for further identification by hybrid-selected translation using the nuclease-treated reticulocyte lysate system.

mRNA recovered from hybridization to DNA of one clone, pWB1, markedly stimulated incorporation of [^{3}H]valine into protein, which was immunoprecipitable with affinity-purified antielastin antibody. When the immunoprecipitated product was subjected to polyacrylamide gel electrophoresis, a major peak of 70,000 daltons characteristic of tropoelastin was observed. Restriction enzyme analysis of the recombinant plasmid demonstrated that the insert was approximately 240 bases.

J. Keratin

A method has been described which allows the isolation of undegraded total cellular RNA from living epidermis in a one-step procedure [303]. Epidermal tissue is homogenized and centrifuged through a 5.7 M CsCl cushion. Under appropriate density conditions in the homogenate, large amounts of RNA are sedimented free of DNA and protein to the bottom of the tubes. A prerequisite for the applicability of this method is the preparation of pure epidermis under conditions that efficiently inactivate endogenous RNases. Oligo(dT) chromatography of total RNA yields 3-4% of poly(A) RNA. Translation of poly(A) fractions from both newborn and adult mouse epidermis in a cell-free reticulocyte system leads to the synthesis of proteins with electrophoretic mobilities identical to those of native keratin polypeptides of the corresponding tissue. The CNBr cleavage pattern of these purified proteins is identical to that of native keratins. There is no evidence for the synthesis of precursors of these proteins.

K. Ovalbumin

Ovalbumin is an egg white protein that accounts for about 60% of the protein synthetic activity of the chicken oviduct, where it has been shown to be under the hormonal control of estrogen and progesterone. Ovalbumin mRNA was purified from hen oviduct by immunoprecipitation [537] of polysomes and oligo(dT)-cellulose chromatography [304] or poly(U)-Sepharose [459]. Two steps were introduced to improve the separation of mRNA and rRNA by oligo(dT)-cellulose. First, aggregates of mRNA and rRNA were dissociated by heating at 65°C for 10 min before chromatography. Second, elution of the mRNA was achieved by stepwise increases in temperature rather than by lowering the ionic strength. Ovalbumin mRNA activity was eluted primarily in RNA fractions eluting between 45 and 55°C. Polyacrylamide gel electrophoresis indicated that the ovalbumin mRNA thus obtained was essentially free of rRNA. The purification of ovalbumin mRNA on DNA-cellulose has been reported [307]. Purified ovalbumin mRNA was employed to isolate the gene coding for ovalbumin from total chick DNA by hybridization [305,306,553-555]. Using mRNA covalently linked onto phosphocellulose, the coding strand of the ovalbumin gene was enriched 10,000-fold from sheared chick DNA. This gene sequence was further purified by sulfhydryl-Sepharose column chromatography after hybridization with an excess of mercurated mRNA. The overall purification of the ovalbumin gene from total chick DNA was thus 180,000-fold, and two of every five DNA molecules in the final preparation contained sequences of the structural ovalbumin gene. There was no apparent degradation of the 5000 nucleotide strands of chick DNA throughout the purification procedure. Since mRNA has a complexity of 1890 nucleotides, the resulting DNA was more than twice the length of mRNA and should contain DNA sequences tandemly linked to the structural ovalbumin gene.

L. Casein

Purification of casein mRNAs from lactating rat mammary gland RNA has been accomplished by Sepharose 4B chromatography, preparative agarose-urea gel electrophoresis, and chromatography of poly(A) mRNA on oligo(dT)-cellulose [308,570]. The separation of the individual casein mRNAs into discrete molecular species free of apparent rRNA contaminants was facilitated by the use of denaturing conditions either prior to or during each of the fractionation procedures. Two casein mRNA fractions were isolated. Under optimal translation conditions for casein mRNA, at least 90% of the released protein synthesized in response to the 15S casein mRNA was specifically immunoprecipitable, representing a 178-fold purification compared with the initial RNA extract. Using these techniques a comparable purification was also obtained for a 15S mouse casein mRNA fraction.

Specific polysomes synthesizing α_S-casein and β-casein were immunoprecipitated from total polysomes of the lactating ewe mammary gland [309]. The polysome-anticasein complex was immunoprecipitated by anti-immunoglobulins; 22%, 32%, and 10% of polysomes were immunoprecipitated with saturating amounts of anti-α_S-casein, anti-β-casein, and anti-x-casein, respectively. Poly(U)-Sepharose chromatography of the immunoprecipitated RNAs permitted isolation of the corresponding poly(A) RNA, which migrated as one major band in polyacrylamide gel electrophoresis. As judged by the contamination with messenger activity for one of the other caseins, the purity of the mRNA for α_s-casein as well as for β-casein was estimated to be between 75 and 80%.

M. Albumin

Rat liver albumin mRNA has been purified [571] to apparent homogeneity by means of polysome immunoprecipitation and poly(U)-Sepharose chromatography [310,572]. The albumin-polysome immunoprecipitate was dissolved in detergent and the polysomal RNA was separated from protein by sucrose gradient centrifugation. Albumin mRNA was then separated from rRNA by chromatography through the binding of poly(U)-Sepharose to the polyadenylate 3' terminus of the mRNA. It coded for the translation of authentic liver albumin when added to a heterologous protein-synthesizing cell-free system derived from either rabbit reticulocyte lysates or wheat germ extracts. Translation analysis in reticulocyte lysates indicated that albumin polysomes were purified approximately ninefold from total liver polysomes, and that albumin mRNA was purified approximately 74-fold from albumin polysomal RNA.

Mouse albumin mRNA was isolated by direct immunoprecipitation of albumin-synthesizing polysomes and oligo(dT)-cellulose chromatography of albumin polysomal RNA [486]. Data from the hybridization analyses and cell-free translation studies indicate that the albumin mRNA is about 80 to 85% pure.

Guinea pig liver albumin mRNA had been purified by means of polysomal immunoprecipitate chromatography. Specific mRNA-synthesizing albumin was obtained through oligo(dT)-cellulose chromatography [487]. The translation assay in wheat germ cell-free system has shown that about 90% translated products can form immunoprecipitate with specific antibody against guinea pig albumin.

α-Lactalbumin mRNA was partially purified from RNA extracted from 3- to 5-day lactating rat mammary glands on a poly(U)-Sepharose column followed by sucrose gradient centrifugation [311]. α-Lactalbumin mRNA activity was assayed in wheat germ cell-free translational system by immunoprecipitation of the in vitro synthesized protein using specific antiserum prepared against purified rat α-lactalbumin. In the purified mRNA preparation α-lactalbumin mRNA activity comprised approximately 85% of the total mRNA activity.

The mRNA for albumin was isolated from the liver of male frogs [312]. Purification was achieved using oligo(dT)-cellulose chromatography and sucrose gradient centrifugation. As judged by translational activity in a cell-free protein-synthesizing system derived from rabbit reticulocytes, albumin mRNA was enriched 259-fold compared to the total RNA of the liver cells. Translation in vitro yielded a product which is immunoprecipitable with anti-(frog albumin). cDNA was synthesized using reverse transcriptase and, as a template, purified albumin mRNA.

N. Zein

Zein mRNA was isolated from membrane-bound polyribosomes of developing maize kernels by oligo(dT)-cellulose chromatography [313]. Translation of the mRNA in vitro yielded protein similar to native zein. It synthesized zein components of both 21,800 and 19,000 molecular weight in the wheat germ cell-free protein-synthesis system. The possibility of a polycistronic mRNA or two mRNAs of similar molecular weight was considered.

O. Storage Proteins

mRNAs in developing cotyledons of *Pisum. sativum* (L) coding for the major storage proteins vicillin and legumin have been partially purified and characterized [314]. Both vicillin and legumin precursor subunits were translated in the reticulocyte lysate cell-free system when driven by 18S poly(A)$^+$ RNA. Total poly(A)$^+$ RNA, purified twice by oligo(dT)-cellulose chromatography, migrated on dissociating gels on electrophoresis as a polydisperse peak. This mRNA preparation was used as a template for the synthesis of single- and double-stranded cDNAs. Analysis of single-stranded cDNA on dissociating gels showed a polydisperse profile, with three major components. The double-stranded cDNA preparation contained some contaminating single strands, some partially double-stranded molecules and some molecules which were primarily single strands protected by partial second strands. Restriction endonucleases cut the double-stranded cDNA preparation into several discrete fragments.

P. Vitellogenin

Vitellogenin is a 240,000-dalton protein that can be induced, by estrogen treatment of egg-laying animals, to become the major protein synthesized by the male liver. The isolation of 29S vitellogenin mRNA has been particularly difficult due to the high endogenous ribonuclease content of amphibian and avian livers [315,574]. RNA extracted from livers of estradiol-17β-treated

X. laevis directs the synthesis of the entire 200,000-dalton vitellogenin monomer in a cell-free protein-synthesizing system derived from rabbit reticulocytes. The in vitro product and [^{14}C]-vitellogenin comigrate on electrophoresis in sodium dodecyl sulfate-polyacrylamide gels and they exhibit identical immunoprecipitation curves. *X. laevis* vitellogenin mRNA can be purified approximately 60-fold from cell RNA by poly(U)-Sepharose chromatography and therefore appears to contain a poly(A) sequence. Vitellogenin mRNA and vitellogenin synthesis in vivo could not be detected in unstimulated male *X. laevis*. Vitellogenin synthesis is maximal 12 days after estradiol-17β administration, when it comprises about 70% of cell protein synthesis. The level of vitellogenin mRNA and the intracellular rate of vitellogenin synthesis exhibit a close correspondence from 4 to 16 days after administration of estradiol-17β. Estrogen-treated rooster livers have been homogenized in 8 M guanidine·HCl containing 1 mM dithiothreitol at pH 5.0 to achieve rapid and effective denaturation of RNase. Poly(U)-Sephadex G10 was found to yield improved recoveries of the large poly(A) mRNA, compared to oligo(dT)-cellulose or poly(U)-Sepharose 4B [316].

Q. S-100 Protein

Poly(A) polysomal RNA was prepared from rabbit cerebral hemispheres using phenol extraction and chromatography on oligo(dT)-cellulose [317]. This RNA directed the synthesis of the brain-specific S-100 protein in cell-free extracts from wheat embryo. S-100 protein was absent from the products of endogenous incorporation and from a reaction programmed with kidney mRNA. These results suggest that S-100 protein mRNA contains a poly(A) sequence and rule out the necessity of a brain-specific factor for translation of S-100 protein mRNA.

R. Fibronectin

Fibronectin mRNA has been partially purified by guanidine extraction, oligo(dT)-cellulose chromatography, and sucrose density gradient centrifugation [318]. A fraction was obtained that programs a wheat germ in vitro translation system to synthesize a polypeptide species which co-electrophoreses with fibronectin in sodium dodecyl sulfate-polyacrylamide gels and which is immunoprecipitated with affinity-purified fibronectin-specific IgG.

S. Immune Interferone

A lysate of *Staphylococcus aureus* induced high-titered mouse immune interferon (MuIFN-γ) in spleen cell cultures [319]. mRNA was extracted from such cells, purified by oligo(dT)-cellulose chromatography and sucrose gradient centrifugation, and injected in *X. laevis* oocytes. Fractions containing RNA of 15S to 16S gave rise to IFN activity that was characterized as IFN-γ by virtue of its acid lability, neutralization by antiserum to partially purified MuIFN-γ, and lack of neutralization by antiserum to NDV-induced L-cell IFN.

Suspension cultures of total and lymphocyte-enriched peripheral white blood cells were used as a source for the preparation of mRNA for human γ-interferon [320]. The cells were lysed and total RNA was extracted by the phenol method. A poly(A) RNA fraction was isolated by chromatography on oligo(dT)-cellulose and further purified on a preparative sucrose gradient. The RNA preparations were translated by oocytes into (a) protein(s) that showed biological activity in the interferon assay.

V. mRNA OF ENZYMES

A. Phosphoenolpyruvate Carboxykinase

Phosphoenolpyruvate carboxykinase is one of the key regulatory enzymes in gluconeogenesis. mRNA was extracted from the livers of starved rats and incubated in a heterologous cell-free protein-synthesizing system derived from rabbit reticulocytes [321]. The presence of newly synthesized phosphoenolpyruvate carboxykinase was detected by immunoprecipitation with a specific antibody to the enzyme and analysis by sodium dodecyl sulfate-polyacrylamide gel electrophoresis. The presence of poly(A) in phosphoenolpyruvate carboxykinase mRNA was suggested by retention of mRNA activity on oligo(dT)-cellulose column. It was concluded that the cell-free synthesis of phosphoenolpyruvate carboxykinase can serve as a bioassay for intracellular phosphoenolpyruvate carboxykinase mRNA.

The mRNA was also partially purified from the liver of cyclic-AMP-treated rats by a procedure involving multiple oligo(dT)-cellulose chromatographies and sucrose gradient fractionations [322]. The purification was monitored by translational assay using a wheat germ extract. Relative to RNA bound once to oligo-(dT)-cellulose, the final material was enriched 20-fold in template activity for phosphoenolpyruvate carboxykinase synthesis. With this RNA preparation, cell-free enzyme synthesis amounted to 5% of total RNA-directed protein synthesis. The apparent sedimentation coefficient of phosphoenolpyruvate carboxykinase mRNA in sucrose gradients was between 20S and 22S, corresponding to an average molecular weight of 0.93×10^6. By formamide/polyacrylamide gel electrophoresis the molecular weight of the enzyme mRNA was estimated at between 0.91×10^6 and 1.12×10^6. From these estimates it was concluded that considerable noncoding sequence(s) are present in the mRNA. Approximately 20% of the enzyme mRNA in rat liver failed to bind to oligo(dT)-cellulose, presumably because of the absence of a poly(A) segment. The translation of phosphoenolpyruvate carboxykinase mRNA by the wheat germ extract was inhibited in the presence of 7-methylguanosine-5'-phosphate. The enzyme mRNA appears therefore to have a "cap" at the 5' end.

B. Large Subunit Ribulose Bisphosphate Carboxylase

RNA from wild-type *Euglena*, aplastidic mutant cells, and purified chloroplasts was chromatographed on oligo(dT)-cellulose [323]. The poly$(A)^+$ and poly$(A)^-$ fractions from each were then tested in a cell-free protein-synthesizing system from wheat germ for their specific ability to stimulate synthesis of large-subunit ribulosebiphosphate carboxylase [EC 4.1.1.39; 3-phospho-D-glycerate carboxylyase (dimerizing)]. The large-subunit polypeptide (59,000 molecular weight) was identified by two-dimensional electrophoresis involving isoelectric focusing and size separation on polyacrylamide gels. Template activity for the large subunit was detected in wild-type cells but not in aplastidic mutant strains; it was highly enriched in the isolated chloroplasts. This messenger was present only in the poly$(A)^-$ RNA fraction, where it constituted the most prominent template species of chloroplast RNA. The large subunit message was freed of considerable non-mRNA contamination and localized to the 10S-20S fraction by sucrose gradient centrifugation.

C. Tyrosine Aminotransferase

A study has been made of the characteristics of an mRNA which has the ability to translate tyrosine aminotransferase (TAT) [324]. Polysomes were incubated with Sepharose, containing covalently bound antibodies to TAT, which had been treated with rat plasma. The polysomes that had absorbed onto the Sepharose were subsequently eluted with EDTA-containing buffer. The eluate was treated with a deproteinized mixture of phenol, chloroform, and isoamyl alcohol, and transferred onto poly(U)-Sepharose. RNA sorbed onto the poly(U)-Sepharose was eluted with distilled water. Immunoprecipitation and immunosorption analyses indicated that up to 70% of the peptides synthesized in a cell-free system are associated with the immunosorbent or interact in immunoprecipitation reactions with specific antibodies.

D. Amidating Enzyme

mRNA required for the synthesis of amidating enzyme in the nuclear fraction of γ-irradiated potato buds has been isolated using poly(U)-Sepharose 4B chromatography and sucrose density gradient centrifugation [325]. A cell-free protein-synthesizing system has been obtained from the irradiated potato bud nuclei, capable of in vitro synthesis of this enzyme protein in the presence of mRNA. Unirradiated bud tissue nuclei does not contain the mRNA for this protein. Isolated mRNA can also be translated with a wheat germ system to give the amidating enzyme. Separation of this mRNA on 5-20% sucrose density gradient shows that mRNA amidating enzyme is of the 9S type.

E. Yeast Iso-1-Cytochrome c

The nucleotide sequence of the yeast iso-1-cytochrome c (CYCI) mRNA is presented [326]. The mRNA was enriched by hybridization to cloned CYCI DNA attached to a solid matrix: either nitrocellulose filters or diazobenzyloxymethyl cellulose powder. The sequence of the 5' end of the mRNA was determined by the extension of a CYCI-specific dodecanucleotide primer; the sequence of the 3' end was determined using a decanucleotide d(pT_8-G-A) primer. The CYCI mRNA begins 61 nucleotides 5' to the AUG initiation codon, extends through the coding sequence to 172 to 175 nucleotides 3' to the UAA termination codon, followed by the poly(A) tail. There are no intervening sequences.

F. Swine Pepsinogen

Pepsinogen, the zymogen of pepsin (EC 3.4.23.1) is produced principally in the gastric chief cells of fundic mucosa. It is then secreted and activated autocatalytically to pepsin. Poly(A) RNA was isolated from total RNA of swine gastric mucosa by oligo(dT)-cellulose chromatography, and was translated in a wheat germ cell-free system [327]. Analysis of the translation products by sodium dodecyl sulfate-polyacrylamide gel electrophoresis revealed that pepsinogen was a major translation product and was synthesized as two different molecular forms with apparent molecular weights of 45,000 and 43,000. The pepsinogen translated in the in vitro translation system had no autocatalytic activity. The pepsinogen mRNA was further purified by sucrose density gradient centrifugation, where the mRNA activity for pepsinogen was located around 15S. At this stage, the translation product of the pooled fractions appeared to be almost exclusively pepsinogen. The peptide maps of the pepsinogen which was translated in vitro and digested by α-chymotrypsin and by *S. aureus* V8 protease were nearly identical with the corresponding peptide maps of standard pepsinogen. The double-stranded cDNA that had been synthesized from the partially purified mRNA by reverse transcriptase was cloned in *E. coli* χ 1776. A colony carrying pepsinogen cDNA sequence was identified by in situ colony hybridization using the cDNA synthesized from the partially purified mRNA as a probe and further by a positive hybridization-translation assay. One of the recombinant clones (pSPcA1) had an insert of about 850 bp and the nucleotide sequence analysis revealed that pSPcA1 codes for swine pepsinogen.

G. Aldolase B Subunit

Fructose 1,6-bisphosphate aldolase (EC 4.1.2.13) appears in three isozyme types: aldolase A (muscle type), aldolase B (liver type), and aldolase C (brain type). In addition to these organ- or tissue-specific expressions, the enzyme changes its isozyme pattern during fetal development or carcinogenesis in the same tissue. Aldolase A is predominant in fetal rat liver, while aldolase B is present almost exclusively in adult rat liver. mRNA coding for aldolase B subunit was purified from rat liver polysomes through specific immunoadsorption, oligo(dT)-cellulose chromatography, and sucrose gradient centrifugation [328]. Translation of the purified mRNA in a rabbit reticulocyte system followed by sodium dodecyl sulfate-polyacrylamide gel electrophoresis revealed that almost all of the translation products comigrated with marker aldolase B subunit. About 85% of the proteins synthesized in vitro were identified by the immunochemical analysis as aldolase B subunit. The activity of the mRNA to synthesize aldolase B was increased to 127 times that of poly(A) RNA from total polysomes. The purified mRNA was about 14.5S as estimated by sucrose gradient centrifugation.

H. Aminoacyl-tRNA Synthetase

The purification of mRNAs for yeast arginyl, asparatyl, and valyl α and β subunits of phenylalanyl-tRNA synthetases has been described [488]. The translation products of mRNAs for arginyl- and valyl-tRNA synthetases have the same molecular weight as the purified enzymes; translation of asparatyl-tRNA synthetase mRNA yielded a 68-kilodalton polypeptide (while the purified crystallizable enzyme appears as a 64- to 66-kilodalton doublet). The translation of the mRNAs for α and β phenylalanyl-tRNA synthetase gave polypeptides having the same molecular weight as those obtained from the purified enzyme.

VI. mRNA OF HORMONES

A. Thyroglobulin

Thyroglobulin is a huge multimeric glycoprotein of approximately 600,000 daltons and a sedimentation coefficient of 19S. It is the major product of protein synthesis by the thyroid gland, which after iodination and coupling of a fraction of its tyrosyl residues serves as storage for inactive thyroid hormones. Three poly(A) RNAs (~33S, 16S, and 13S) have been isolated from membrane-bound thyroid polysomes by chromatography on poly(U)-Sepharose [329]. The 33S RNA could also be characterized from thyroglobulin-synthesizing polysomes by hybridization of its poly(A) sequence with tritiated poly(U). When injected into *Xenopus* oocytes, the 33 and 16S RNA promoted the synthesis of immunoprecipitable thyroglobulin.

Poly(U)-Sepharose chromatography of the RNA extracted from a thyroid fraction allows the purification of two RNA fractions amounting each to 1% of the applied material [330] (Fig. 14). The first one is loosely bound to the column, from which it is eluted at 25°C. It is mainly composed of 16S and 12S RNA comprising no poly(A) sequences. This could correspond to mitochondrial rRNA. The second one, which is eluted at 50°C, is poly(A)$^+$ and represents 33S and 17S-18S RNA species. The 33S RNA resists heating at 80°C, suggesting that it is composed of one polynucleotide chain. When injected into *Xenopus* oocytes, the 33S RNA specifically promotes the synthesis of a peptide with an apparent molecular weight of 185,000 and an apparent sedimentation coefficient of 10S. This peptide is immunologically related to thyroglobulin and could represent its main precursor. Under the conditions tested, it does not polymerize spontaneously into 19S

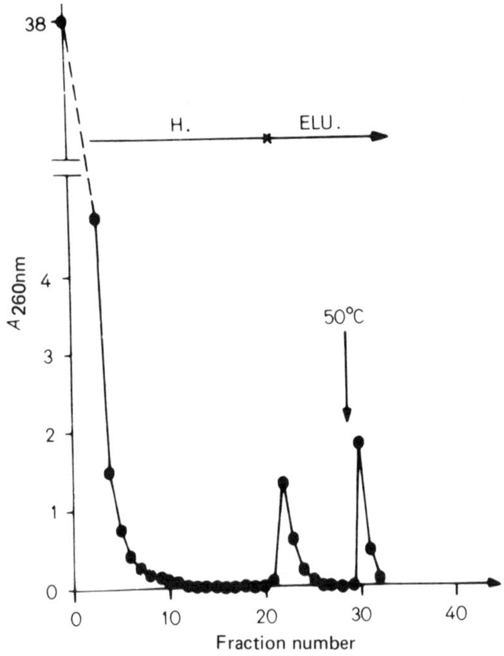

Figure 14 Elution pattern of RNA purified by poly(U)-Sepharose chromatography. One hundred ninety A_{260nm} units of RNA extraction from a thyroid fraction were mixed with 3 ml of poly(U)-Sepharose. Hybridization was allowed to take place in hybridization buffer (H), after which the column was extensively washed at 25°C with the same buffer. Elution was then performed, first by replacing the eluant by the elution buffer (ELU) and then by heating the column to 50°C. (From Ref. 330.)

thyroglobulin, suggesting that assembly of the molecule could require specific, post-translational alterations of the precursor and/or the presence of additional lighter subunits. Chemical identity to the native protein is presented by means of a tryptic peptide analysis [331]. Moreover, the 33S mRNA is shown to contain all the information required for the synthesis of a complete 19S thyroglobulin molecule. Gel filtration in Sepharose indicates that the reduced polypeptide encoded by the 33S mRNA is larger than 210,000 daltons.

Human thyroglobulin mRNA was isolated from Graves' goiters by size selection of total poly(A) RNA in a sucrose gradient [482]. cDNA was synthesized from the 33S RNA. After tailing with dCTP and terminal transferase, the double-stranded cDNA was annealed to pBR322 DNA that had been cleaved at the endonuclease PstI site and tailed with dGTP. The resulting plasmids were used to transform *E. coli* C600 cells and four cloned recombinants were selected. Each plasmid DNA was shown to contain a sequence complementary to human thyroglobulin mRNA by visualization of cDNA·mRNA hybrids by electron microscopy and filter hybridization selection of mRNA directing the synthesis of immunologically related thyroglobulin peptides in the reticulocyte lysate. The four inserted DNA sequences were 1400-1800 bp long, two of them showing an homologous sequence of 1100 bp. Together, the four cloned DNA fragments represented 63% of the 8500 bases of human thyroglobulin mRNA.

B. Preproinsulin

Electrophoretically homogeneous mRNA for rat preproinsulin has been prepared from an islet cell tumor by the use of oligo(dT)-cellulose chromatography and sucrose gradient centrifugation [332]. The results indicate that the synthesis of preproinsulin is directed by a monocistronic mRNA 600 nucleotides in length and requires about 55% of the molecule's coding capacity.

C. Gastrin

Three oligodeoxynucleotides have been chemically synthesized and used as hybridization probes to detect gastrin-specific mRNA within an heterogeneous population of RNA molecules [333,565]. Using these oligonucleotides, 0.2 fmol of gastrin mRNA can be detected per microgram of poly(A) RNA from hog antrums. The [^{32}P]-labeled oligonucleotides hybridize specifically to RNA covalently coupled to DBM paper. Labeled oligonucleotides also hybridize specifically to RNA separated by gel electrophoresis in the presence of CH_3HgOH and covalently coupled to DBM paper.

D. Bovine Prolactin and Growth Hormone

Methods are described for obtaining and characterizing highly enriched preparations of bovine prolactin and growth hormone mRNAs [334]. Phenol extraction, oligo(dT)-cellulose chromatography, sucrose gradient ultracentrifugation, and polyacrylamide gel electrophoresis results in messenger preparations that are about 92% pure for prolactin mRNA and 78% pure for the growth hormone mRNA species. Translation experiments with 7-methylguanosine, and end labeling of the nucleic acid before and after β oxidation, indicate that both messages possess blocked 5' termini. Poly(A) tracts of 30-160 residues were found at the 3' ends of both purified species. Sizing experiments suggest that both mRNAs contain about 30% more bases than are accounted for by the coding regions and the poly(A) tracts.

Poly(A) RNA from the bovine anterior pituitary has been used as a template for the enzymatic synthesis of double-stranded cDNA [335]. The resulting double-stranded cDNA was inserted into the PstI site of pBR322 and subsequently cloned in *E. coli* χ 1776. A 250 sequence from one prolactin-positive clone was characterized extensively and shown to contain the coding information for amino acids 119-192 of authentic bovine prolactin. The recombinant DNA from this clone was covalently attached to diazotized aminocellulose and used to purify prolactin mRNA from a mixture of mRNAs.

DNA complementary to mRNA coding for bovine prolactin (bPrl) has been cloned [336]. A recombinant plasmid containing bPrl cDNA was identified by hybridization to cloned rat Prl cDNA. It contains cDNA corresponding to the region of the mRNA coding for the carboxy-terminal 101 amino acids of bPrl, as well as 42 nucleotides in the 3' untranslated region of the mRNA. Nucleotide sequencing confirmed the amino acid sequence of this region of bPrl, and permitted the assignment of asparagine or glutamic acid at seven previously equivocal loci. Codon use in bPrl mRNA is comparable to that found in rat and human Prl mRNAs and differs from that in bovine, rat, and human growth hormone mRNAs.

DNA complementary to mRNA coding for bovine growth hormone (bGH) has been cloned [337]. A recombinant plasmid containing bGH cDNA was identified by hybridization to cloned rat growth hormone cDNA. It contains the entire coding and 3' untranslated regions and 31 bases in the 5' untranslated region. Nucleotide sequence analysis determined the sequence of the 26-amino acid signal peptide and confirmed the published amino acid sequence of the secreted hormone at all but two residues.

VII. mRNA OF PROTAMINE AND HISTONE

One of the smallest mRNAs that has been isolated is protamine mRNA from trout testes. Phenol-chloroform-isoamyl alcohol extraction of the microsomal tissue fraction, chromatography on oligo(dT)-cellulose, and sucrose gradient fractionation have been used to isolate the 6S poly(A) protamine mRNA species [338]. Examination of this material by electrophoresis in polyacrylamide gels containing 8 M urea revealed the presence of four distinct mRNA subclasses [339,543].

In a variety of systems, histone mRNA has been shown to lack poly(A). It was found, however, that in *X. laevis* oocytes, poly(A) mRNA codes for histones, in a wheat germ cell-free system [340]. RNA which directs the synthesis of histones in the cell-free system is retained on oligo(dT)-cellulose, even after denaturation in 80% DMSO at 70°C, thereby demonstrating the covalent attachment of poly(A) sequences to the mRNA [483]. Poly(A) RNA (7S-14S fraction) was also found to code for histones.

Cloned sea urchin (*Strongylocentrotus purpuratus*) histone gene fragments linked to cellulose have been used as hybridization probes to isolate specific histone mRNAs from whole cell and polysomal RNA extracts [341]. RNAs isolated in this manner maintain their biological activity, serving as templates for histone proteins in a wheat germ cell-free protein translation system. In addition, radiolabeled histone-specific RNA purified from cleavage-stage sea urchin embryos, pulse labeled for short periods of time, and analyzed on denaturing polyacrylamide gels was the same size as mature histone mRNAs.

The use of a synthetic primer to select a cDNA recombinant clone containing H5 coding sequences has been described [342]. The strategy used was as follows. Prepare oligo(dT)-cellulose-bound mRNA from chicken reticulocytes and select 11S-18S material from sucrose gradients. Use this RNA fraction both to prepare a cDNA library and as a template for H5-specific cDNA synthesis using a synthetic primer. Screen out most globin cDNA recombinants with oligo(dT)-primed globin cDNA. Search for H5 recombinants using H5-specific cDNA and verify the identity by DNA sequencing. The isolation of an H5 cDNA recombinant clone is an initial step in the study of H5 genes and their relationship to H1 and core histone genes.

9

Isolation of Proteins

The various types of immobilized ligands, such as nucleobases, nucleosides, mononucleotides, and their derivatives, have been used to bind selectively a number of different enzyme groups and it has been possible to observe general patterns in the binding and elution characteristics of the various ligand types [3]. Because a template function is based on a certain chain length of the immobilized ligand isolations of enzymes using immobilized nucleic acid components, however, may be better reckoned to the field of general affinity chromatography and are apart from some examples not reviewed here.

We have reported in earlier chapters on all preparations of different proteins which have used DNA, RNA, tRNA, and oligonucleotides immobilized to various polymer supports. These proteins are of immense interest to both the molecular biologist and the biochemist. Among them are several DNA polymerases, the RNA polymerases, DNA regulatory proteins, RNA-dependent DNA polymerase, repressor proteins, ribosomal proteins, tRNA synthetases, and viral DNA- and RNA-binding proteins. Template chromatography has been used successfully in the purification of several proteins. Another application lies in the screening of crude extracts for new types of proteins. A selection of typical approaches is described below.

I. ISOLATION OF ENZYMES INVOLVED IN THE REPLICATION, TRANSCRIPTION, TRANSLATION, OR DEGRADATION OF NUCLEIC ACIDS

Many proteins that operate in association with intracellular DNA recognize purified DNA as a substrate and bind tightly to it at physiological ionic strengths in vitro. When DNA-free, crude extracts are passed through a column consisting of DNA adsorbed onto an inert matrix, these proteins are specifically retained. After an extensive wash to remove unbound proteins, the DNA-binding proteins are recovered in active form by elution with buffers containing either an elevated salt concentration, competing nucleic acids, or specific biological inducers. With an appropriate choice of condition, most adventitious binding of non-DNA-related proteins to DNA can be avoided.

A. Polymerase, Polynucleotide Phosphorylase, Terminal Nucleotidyltransferase, T_4 RNA Ligase, and Topoisomerase

In initial experiments with DNA-cellulose columns the chromatographic behavior of DNA and RNA polymerase enzymatic activities from crude extracts of *E. coli* has been investigated [53,55]. DNA entrapped in polyacrylamide gel particles, which can be used in column chromatography, was investigated for use in the template chromatography of *E. coli* DNA polymerase [73]. The crude enzyme was applied at low ionic strength and eluted at high ionic strength with a 200-fold increase in specific activity on a single passage through the column.

When clarified cell sonicates were chromatographed, major portions of both DNA and RNA polymerase activities are washed through the DNA-cellulose column and are bound to the endogenous DNA of the extract. It appears, therefore, that DNA-free crude extracts are necessary for efficient use of this method. Such extracts are prepared from cells by one of two alternative procedures: (a) pancreatic DNase treatment followed by extensive dialysis against an EDTA-containing buffer to remove divalent cations and inactivate DNase (which would otherwise digest DNA-cellulose), or (b) precipitation of the DNA with polyethylene glycol (PEG) and high salt (to dissociate bound proteins), followed by dialysis of the protein-rich supernatant back to lower ionic strengths. The PEG procedure should be used if chromatography is carried out in the presence of Mg^{2+}. Both the pancreatic DNase and the PEG pass through a DNA-cellulose column; they are therefore readily removed from the fractions of interest.

An alternative procedure for nucleic acid removal that should allow chromatography in the presence of Mg^{2+} is DNA (and RNA) digestion with micrococcal nuclease; since this enzyme requires Ca^{2+}, it can readily be inactivated by use of the specific chelating agent ethylene glycol-bis(aminoethyl ether)tetraacetic acid (EGTA).

Nucleic acid-free extracts of *E. coli* have been analyzed by chromatography on columns of cellulose to which poly(A), poly(U), or poly(C) has been attached by UV irradiation [60]. Proteins are released from the columns by stepwise elution with increasingly higher concentrations of salt, followed by washing with urea. The pattern of protein elution is reproducibly different for each of the homopolymer RNA-cellulose columns used. The RNA polymerase core enzyme binds to poly(C)- and to poly(U)-cellulose columns, and can be purified to 20 to 30% homogeneity in a single step. Ribosomal protein S1 and the termination factor ρ bind very tightly to poly(C)-cellulose, and both can rapidly be purified to homogeneity. Poly(A)-cellulose chromatography allows the isolation of large amounts of an 80,000-dalton protein having an as yet unassigned cellular function. The host factors. required for RNA phage Qβ RNA replication in vitro can also be obtained from poly(A)-cellulose, and chromatography of extracts of phage Qβ-infected *E. coli* on RNA-cellulose columns results in very rapid isolation of the Qβ replicase enzyme.

Agarose gels containing immobilized single-stranded circular DNA from phage fd or denatured calf thymus DNA were investigated [66] for their use in the template chromatography of DNA-binding enzymes for the following reasons: (a) single-stranded DNA is a suitable substrate analog for many DNA-binding enzymes; (b) high concentrations of denatured DNA can be easily immobilized in the porous agarose matrix; and (c) commercially available DNA preparations can be used. The matrix contains little solid material; therefore, unspecific adsorption is expected to be low. Single-stranded DNA-agarose meets all the basic requirements needed for template chromatography of DNA-binding proteins. Single-stranded DNA-agarose columns were used to differentiate between similar enzymatic activities in DNA-free extracts from *E. coli*. These columns select a group of proteins the main part of which are different from those binding to double-stranded DNA columns. Of the soluble *E. coli* proteins, 3% bind to single-stranded DNA-agarose, whereas only 1% binds to double-stranded DNA-cellulose [55]. To demonstrate the resolving power of a DNA-agarose column a DNA-free crude *E. coli* extract was chromatographed and screened for different enzymatic activities.

Preparative purification is described for the following enzymes: *E. coli* DNA polymerase I, DNA polymerase II, RNA polymerase, exonuclease III, and T_4 polynucleotide kinase. Enzyme purification was as high as 200-fold, recovery of enzymatic activity was 75-100%. Gradient elution and assays for RNA polymerase, DNA polymerase, nuclease, and polynucleotide ligase activity resulted in a characteristic pattern which was also obtained when the DNA-specific fraction was rechromatographed. The nucleases do not bind very firmly and are eluted first. The main DNAase activity is retained only on single-stranded DNA and not on double-stranded DNA columns, whereas DNA polymerase I and RNA polymerase bind to both resins.

A simple method was presented to isolate purified holoenzyme of *E. coli* RNA polymerase [344]. When a purified enzyme preparation was chromatographed on a DNA-cellulose column [53], holoenzyme was separated from core enzyme. Thus holoenzyme was eluted at 0.15 M KCl and core enzyme at 0.25 M KCl.

A drawback of DNA-column material used so far has limited their application. In DNA-cellulose [55,61], DNA-polyacrylamide [73], and DNA-agarose [66] the nucleic acid is not bound covalently, imposing a restriction on the ionic strength of the buffer used to avoid a loss of DNA. Since Sepharose was found to possess advantageous properties of high capacity and good flow rates when used in columns, the availability of Sepharose-DNA and Sepharose-RNA matrices would be a useful tool for the isolation and purification of proteins which bind to nucleic acids. DNA has been covalently linked to insoluble matrices of Sepharose in high yield using CNBr activation. Both double-stranded and single-stranded DNA have been coupled [74,77]. The DNA-Sepharose is stable to high temperature, prolonged storage, and in the case of single-stranded DNA can be washed with NaOH to destroy nuclease activity and to release any digested oligonucleotides or mononucleotides.

The DNA-Sepharose has been used for the template chromatography of various enzymes (*E. coli* DNA polymerase I and RNA polymerase) from crude extracts or after initial purification steps, resulting in high yields and degrees of purification [77].

The complex formed through immobilization of bacteriophage f2 RNA on cellulose UV irradiation served in the chromatographic purification of the poly(G) polymerase obtained from f2-infected *E. coli* cells [62].

A DNA polymerase from *Micrococcus luteus* has been isolated using DNA-cellulose chromatography. This approach has permitted the isolation of a highly purified enzyme from a partially purified preparation in which the polymerase is only a minor con stituent. A considerably disadvantageous complication to this procedure is that the cellulose-bound DNA is heavily irradiated with UV light. An endonucleolytic activity specific for UV irradiated DNA was found to be present in the polymerase preparations, although heavily irradiated DNA itself does not serve as a primer for the *M. luteus* DNA polymerase [61].

A high-molecular-weight membrane-bound DNA polymerase from the mouse myeloma, MOPC-104 E, has been purified extensively. This enzyme, which is readily distinguishable from other myeloma enzymes analogous to the recognized forms of cellular DNA polymerase, is designated DNA polymerase III. DNA polymerase III from MOPC-104 E after being homogenized was solubilized and then purified using a series of ion-exchange chromatographic procedures follwed by DNA-cellulose chromatography and glycerol gradient centrifugation; the enzyme activity was purified as much as 18,000-fold. In the final stages of the purification, DNA polymerase III possessed no detectable RNA polymerase activity toward DNA or single- and double-stranded RNA [343].

The passage of a crude soluble enzyme preparation from sonicated HeLa cells through a HeLa DNA-Sepharose column resulted in a enzymatic activity which is purified approximately threefold over the crude sonicated supernatant fluid. The capacity by which the DNA-Sepharose column can absorb the HeLa DNA polymerase is 30- to 60-fold better than a corresponding DNA-cellulose column when calculated per unit of DNA fixed to the matrix [74]. In addition, the Sepharose columns exhibit faster flow rates than corresponding cellulose columns, and like other columns of this type, they can be used repeatedly.

The preparation of CM-cellulose containing similarly high amounts of covalently attached DNA has been reported and its successful application in the purification of DNA-dependent DNA polymerases from yeast nuclei and mitochondria has been described [100]. The advantages of the DNA-CM-cellulose are a rapid and simple preparation procedure, a high yield of bound DNA, especially at input amounts around 10 mg DNA per gram of matrix, and the possibility to link native as well as denatured DNA to the cellulose. Columns can be used repeatedly and allow high flow rates.

Poly(dC)-dependent dGMP incorporating activity without a primer molecule and its stimulating factor were partially purified by successive column chromatographies [544]. Polymerase activity that was highly dependent on the stimulating factor was separated from similar activity that was not stimulated by this factor by native DNA-cellulose column chromatography. The factor stimulating activity was strictly dependent on dGTP as substrate and incorporated dGMP into the 3'-OH terminus of poly(dC). No terminal deoxynucleotidyl transferase activity was detected in the preparation. The activity also responded to heat-denatured calf thymus DNA and poly(dT) as template, although to a lesser extent. The activity was inhibited by dideoxy-GTP and N-ethylmaleimide, and was decreased significantly by aphidicolin and β-lapachon.

The isolation of RNA-dependent DNA polymerase from RNA tumor viruses has been described [347] and the polymerase from Rauscher murine leukemia virus has been purified by a one-step enrichment procedure on columns of $(dT)_{12-18}$-cellulose [348].

Poly(C) covalently linked to CNBr-agarose is an effective matrix for the RNA-dependent DNA polymerase from avian myeloblastosis virus [349]. The DNA polymerase isolated from crude, detergent-disrupted virions by a single pass through columns of poly(C)-agarose appears nearly homogeneous. Complete recovery of input enzymatic activity was obtained.

Foot-and-mouth disease virus RNA polymerase was purified from the polyethylene glycol-treated supernatant of infected cell media by a combination of ion-exchange chromatography, membrane molecular filtration, and template chromatography [507].

The purification of *B. subtilis* RNA polymerase, based on mild lysis of cells, low-speed centrifugation, gel filtration, DEAE-Sephadex chromatography, and template chromatography on DNA-cellulose [55], yields three forms of enzyme, A, B, and C [345]. As revealed by sodium dodecyl sulfate-gel electrophoresis, enzyme A has the subunit structure of core polymerase plus some small polypeptides. Its catalytic properties are similar to those of core polymerase. Enzyme B has the composition of core polymerase. Enzymes A and B can both be stimulated by the addition of σ factor. Enzyme C has the holoenzyme composition. The patterns of sensitivity of the three forms of enzyme toward KCl are very different. The transcripts of the three enzymes on phage SPPI DNA template have been analyzed by hybridization to the separated strands. Only enzyme C selectively transcribed the H strands.

Poly(ADP-ribose) polymerase with high specific activity was obtained from Ehrlich ascites tumor cells [346] by extraction of nuclei with 175 mM potassium phosphate, followed by chromatography on DNA-agarose [66]. Electrophoretic analysis indicated that the preparation contained two proteins, one of which was shown to catalyze the synthesis of poly(ADP-ribose). The synthesis was inhibited by nicotinamide and thymidine, and stimulated by DNA. The addition of histones gave inhibition of the synthesis, unless DNA was present in the reaction mixture. α and α,β DNA polymerases of avian myeloblastosis virus bound effectively to poly(A)·oligo(dT)-cellulose [350]. α DNA polymerase did not bind to oligo(dT)-cellulose, poly(A)-cellulose, or to cellulose. α,β bound to oligo(dT)-cellulose and cellulose at the same extent, indicating that this enzyme did not bind specifically to the oligo(dT) moiety only. However, α,β bound to poly(A)-cellulose two to three times better than to cellulose itself, showing that α,β could bind to poly(A) without a primer. α,β DNA polymerase also bound to poly(C)-cellulose, whereas α did not. These data show that the α DNA polymerase is defective in binding to nucleic acids if the β subunit is not present. Data are presented which demonstrate that the α,β DNA polymerase bound tighter than the α DNA polymerase to poly(A)·oligo(dT)-cellulose and to calf thymus DNA-cellulose, suggesting that the β subunit or, at least part of it, is responsible for this tighter binding. In addition, α,β DNA polymerase is able to transcribe reversibly avian myeloblastosis virus 70S RNA about fivefold faster than α DNA polymerase in the presence of Mg^{2+} and equally efficient in the presence of Mn^{2+}. α DNA polymerase transcribed 9S globin mRNA slightly better than did α,β with either metal ion.

DNA-binding activity of the extracted RNA polymerase II by chromatography of native DNA that was covalently coupled to Sephadex G-25 has been examined [489]. The free RNA polymerase II of the nuclear soluble fraction revealed very low binding activity compared with the enzyme extracted from the chromatin-bound fraction. The nuclear-soluble RNA polymerases, once purified by DEAE-Sephadex A-25 chromatography, acquired DNA-binding activities. The weak binding of the nuclear soluble enzymes was due to association of heat-stable protein(s).

The isolation of polynucleotide phosphorylase from *E. coli* using poly(A)-Sepharose is reported [352]. p-Aminophenyl oligo-(dT)-Sepharose has been used for the purification of polynucleotide phosphorylase from both *E. coli* and *B. stearothermophilus* [41]. *E. coli* polynucleotide phosphorylase was also retarded on DNA-agarose [353].

E. coli native polynucleotide phosphorylase and *E. coli* RNA polymerase (both holoenzyme and the core enzyme) form a complex with blue dextran or polynucleotide immobilized on agarose [354]. The use of these columns allows us to separate native enzyme from its degraded form and to separate σ subunit from the core enzyme of RNA polymerase.

Terminal deoxynucleotidyltransferase is an unusual DNA polymerase because it synthesizes deoxynucleotidyl chains without template instruction. This enzyme is normally found associated with thymus cells, certain bone marrow cells, as well as leukocytes from patients with acute lymphoblastic leukemia and chronic myelocytic leukemia in blast crisis. By using a 35-70% $(NH_4)_2SO_4$ cut, a Sephacryl S200 column, and an oligo(dT)-cellulose column, terminal deoxynucleotidyltransferase has been purified from calf thymus cells to a specific activity of more than 8500 units per milligram of protein [355,490] (Fig. 15). The terminal deoxynucleotidyltransferase purified by this method contains no detectable DNA-dependent DNA polymerase or endonuclease activities. Furthermore, sodium dodecyl sulfate-polyacrylamide gel electrophoresis showed that the enzyme appears to be homogeneous, with two polypeptides corresponding to the two subunits α (10,000) and β (23,000) of terminal deoxynucleotidyltransferase.

A new method is reported for purifying the RNA ligase found in phage T4-infected *E. coli* cells [491]. It was found that RNA ligase could bind to 2',5'-ADP-Sepharose in the presence of Mg^{2+}. 2',5'-ADP-Sepharose chromatography was introduced at the last step of the purification procedure. The purified RNA ligase preparation gave one major band on sodium dodecyl sulfate/acrylamide gel electrophoresis and was estimated to be 80% pure. The purified enzyme had no detectable RNAase activity and was suitable for the joining of short oligoribonucleotides.

An enzyme that catalyzes the addition of a single UMP residue from UTP to the 3' end of an RNA primer, referred to as terminal uridylyltransferase (TUT), has been purified extensively from the membrane fraction of *Vigna unguiculata* leaves [356]. The purification procedure involved solubilization by cation depletion, DEAE-Sepharose CL-6B column chromatography, chromatography on poly(U)-Sepharose 4B, and glycerol gradient centrifugation. The molecular weight of the native enzyme was about 50,000. TUT may be a feasible tool in RNA sequencing, as it enables the specific 3'-terminal labeling of RNA molecules.

Activity of DNA topoisomerase I has been characterized in extracts of mitochondria purified from *X. laevis* oocytes [357]. The enzyme has been purified by DEAE-cellulose, phosphocellulose, and DNA-cellulose chromatography and its properties compared to those of its nuclear-cytosolic counterpart. The mitochondrial enzyme is able to remove positive superhelical turns and possibly acts as a swivelase during replication of mitochondrial DNA.

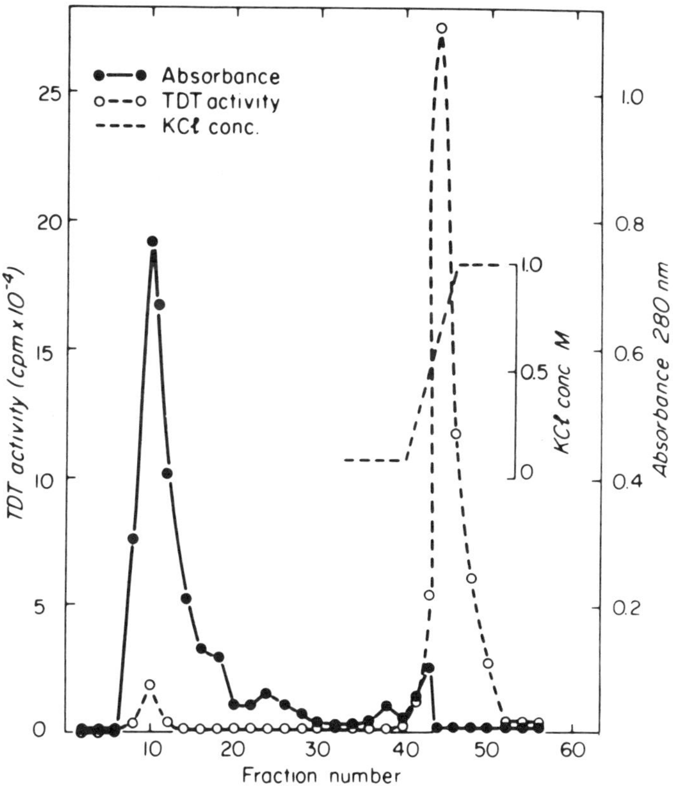

Figure 15 Purification of terminal deoxynucleotidyltransferase (TDT) by an oligo(dT)$_{12-18}$-cellulose column. The column was washed with 35 ml of buffer containing 40 mM KCl, 1 mM dithiothreitol, 10 mM K_2PO_4 (pH 7.1), and 20% glycerol. The enzyme activity was eluted with 1 M KCl in the same buffer. o——o, TDT activity in 25 μl; •——•, absorbance at 280 nm. (From Ref. 355.)

An early event in the initiation of adenovirus DNA replication is the formation of a covalent complex between the 87,000-dalton adenovirus terminal protein precursor and 5'-dCMP(pTP-dCMP complex) [500]. Nuclear extracts prepared from adenovirus-infected HeLa cells catalyzed complex formation in the presence of ATP, Mg^{2+}, and adenovirus DNA-protein complex but were not active when pronase-treated DNA was used as template. The activity has been partially purified by chromatography on denatured

DNA-cellulose and used to examine whether the 55,000-dalton terminal protein on adenovirus DNA is required for pTP-dCMP complex formation.

B. Aminoacyl-tRNA Synthetases

Immobilized tRNA has been employed in a number of studies on the isolation of specific aminoacyl-tRNA synthetases. Chromatographic purifications have been effected by $tRNA^{val}$ or $tRNA^{lys}$ bound to a polyacrylhydrazide agar mixture [358].

Template chromatography was performed on a $tRNA^{leu}$-Sepharose [359]. The matrix, besides the enzyme, retained considerable amounts of other proteins. The specificity of the matrix, therefore, was not as strict as was expected from the specificity of the aminoacylating reaction. A specific purification of leucyl-tRNA synthetase (75-fold), however, was accomplished by choosing suitable conditions for the eluting step. The isolation of yeast phenylalanyl-tRNA synthetase in a completely pure state has been described using $tRNA^{phe}$ immobilized on a hydrazinyl-Sepharose [109]. Chromatography of yeast isoleucyl-, leucyl-, phenylalanyl-, tyrosyl-, and valyl-tRNA synthetase on a column of $tRNA^{phe}$-Sepharose has shown that at pH 5.5 and ionic strength 0.07, these synthetases are bound to the $tRNA^{phe}$ moiety. In contrast to the other synthetases, the leucyl enzyme is not retained on the column either at pH 5.5 and ionic strength 0.17 or at pH 7.5 and ionic strength 0.07. At pH 7.5 and ionic strength 0.17 only the phenylalanyl-and tyrosyl-tRNA synthetases are bound to $tRNA^{phe}$. The ability of synthetases to interact with $tRNA^{phe}$ bound to Sepharose under a variety of conditions follows the order Phe > Tyr > Ile > Val > Leu. Unfractionated tRNA present in the enzyme preparation reduces the interactions between $tRNA^{phe}$ bound to Sepharose and noncognate synthetases [361]. Arginyl-tRNA and asparagyl-tRNA synthetase form yeast are purified in one step using blue dextran or polynucleotide immobilized on agarose [354]. This procedure is particularly useful when the amount of cell extract is very small.

Phenylalanyl-tRNA synthetase was purified from soybean cotyledons [362]. Although the cognate tRNA was used as the ligand, the method used in the preparation of the column did not involve the prior separation of the cognate tRNA from the noncognate tRNA. The enzyme was first purified on a column having noncognate tRNA and then on a column containing cognate tRNA. The properties of the purified enzyme were studied and compared with other aminoacyl tRNA synthetases of the same tissue and also with the phenylalanyl-tRNA synthetase of other plant tissues. Valyl-tRNA synthetase from the human pathogenic strain *Mycobacterium tuberculosis* $H_{37}R_V$ has been purified to homogeneity by chromatography on valyl-aminohexyl Sepharose as the final step of purification [363].

Template chromatography of *E. coli* methionyl-tRNA synthetase can be carried out on agarose-hexyl-AMP in the presence of L-methioninol [364,365]. Several other bacterial aminoacyl-tRNA synthetases have been purified and examined for their behavior on the gel in the presence of the corresponding amino acid analog, which does not have a carboxylate group. Conditions ensuring biospecific retentions on the gel of *E. coli* isoleucyl-, valyl-, tryptophanyl- and tyrosyl-tRNA synthetases and of *Bacillus stearothermophilus* methionyl- and tyrosyl-tRNA synthetases have been found. These enzymes could be recovered from the chromatographic medium by addition of excess L-amino acid to the equilibration buffer containing the analog. Elution under these conditions results from the amino acid antagonizing the nucleotide binding to the enzyme. On the other hand, *E. coli* phenylalanyl-, leucyl-, lysyl-, histidyl-, and cysteinyl-tRNA synthetases could not interact with the agarose hexyl-AMP gel. In the case of *E. coli* threonyl-tRNA synthetase, a nonspecific retention was obtained which depended on the presence of the analog 1-amino-2-propanol as well as on that of the noncognate phenylalaninol or methioninol.

E. coli isoleucyl-tRNA can be bound to bromoacetamidobutyl-agarose, and the resulting complex serves as a chromatographic system for the purification of the corresponding tRNA synthetase [360].

A method of chromatography on Sepharose with immobilized tRNA in the presence of urea was developed for separating the subunits of phenylalanyl-tRNA synthetase from *E. coli* MRE-600 (subunit structure $\alpha_2\beta_2$) [492]. Specific binding of large β subunits of the enzyme on immobilized tRNA testifies to the localization of the tRNA-binding center on the β subunit of phenylalanyl-tRNA synthetase. Separately, α and β subunits of the enzyme exhibit no catalytic activity. Incubation of the mixture of α and β subunits in conditions leading to reassociation of the oligomeric structure results in restoration of catalytic activity of the enzyme. In the presence of urea, resin with immobilized analogs of ATP binds α and β subunits of the enzyme. This testifies to the presence of nucleotide-binding sites on both subunits.

C. tRNA (Adenine-1-)-Methyltransferase

tRNA (adenine-1-)-methyltransferase was isolated from the extreme thermophile *Thermus flavus*, strain 71. It was purified about 2000-fold by ammonium sulfate fractionation and chromatography on tRNA bound to aminohydroxybutyl cellulose via its oxidized 3' end [502] (Fig. 16). The purified protein preparation is free of nuclease and aminoacyl-tRNA synthetase activity and contains no more than 4% of tRNA (guanine-7-)-methyltransferase

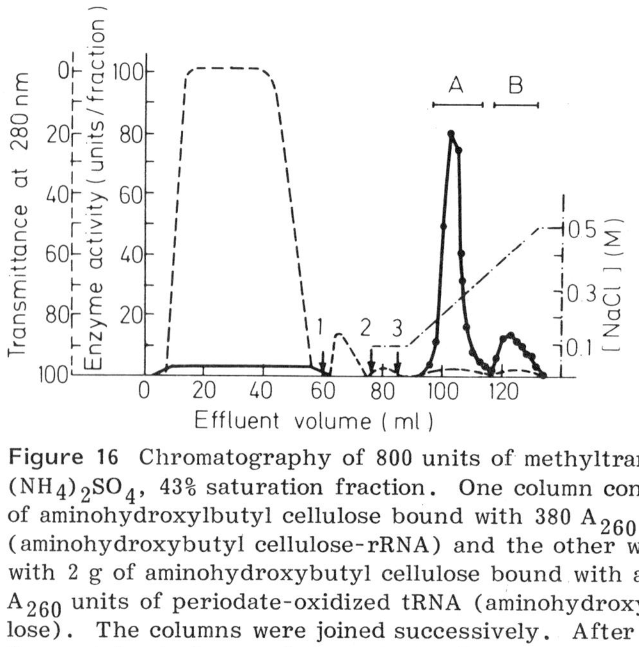

Figure 16 Chromatography of 800 units of methyltransferase $(NH_4)_2SO_4$, 43% saturation fraction. One column contained 2 g of aminohydroxylbutyl cellulose bound with 380 A_{260} units of rRNA (aminohydroxybutyl cellulose-rRNA) and the other was packed with 2 g of aminohydroxybutyl cellulose bound with about 320 A_{260} units of periodate-oxidized tRNA (aminohydroxybutyl cellulose). The columns were joined successively. After the bulk of the proteins had passed through the first column, it was detached, and aminohydroxybutyl cellulose oxidized tRNA was further washed with buffer B, with buffer C (arrow 1), and with buffer C containing 0.1 M NaCl (arrow 2). Buffer B: 0.01 M Tris/HCl pH 6.9, 0.02 M $MgCl_2$, 0.3 M CH_3COONH_4, 0.01 M 2-mercaptoethanol; buffer C: 0.01 M Tris/HCl pH8.2, 0.002 M $MgCl_2$, 0.01 M 2-mercaptoethanol. Methyltransferases were eluted with buffer C using NaCl gradient of 0.1-0.5 M (arrow 3). The flow rate was 10 ml/hr and the fraction size 2 ml. ---, Transmittance at 280 nm; -●-●-●, NaCl concentration; ●——●, tRNA methylase activity. (From Ref. 502.)

activity. The only activity of the enzyme is to methylate A^{58} in the T4 loop of tRNA. Out of the eight purified tRNAs examined. only yeast $tRNA^{Trp}$ was not utilized as a substrate. The enzyme is highly thermostable. It is most active at 75°C.

D. Nucleases

Template chromatography of bovine pancreatic deoxyribonuclease and hog spleen deoxyribonuclease have been studied [366]. Very different binding of both enzymes to single-stranded calf thymus DNA-Sepharose matrices [74] was observed. The binding of hog spleen deoxyribonuclease to a polynucleotide inhibitor Sepharose matrix appeared to be irreversible under the experimental conditions used.

Purification of RNase T_1 has been achieved using a porous glass-2',(3')-GMP adsorbent. This adsorbent exhibits specific affinity for RNase T_1 and is almost devoid of nonspecific ion-exchange properties [367].

Human liver RNase was purified 3600-fold into an electrophoretically homogeneous state by column chromatography on phosphocellulose, gel filtration, poly(G) chromatography, and heparin affinity chromatography [493]. RNase was a heat- and pH-stable protein, and optimal activity was obtained at pH 7.0.

RNase T_2 bound to 5'-adenylate-aminohexyl-Sepharose 4B, specifically at pH 4.5 [23]. The colorless enzyme was eluted only by the simultaneous addition of 2'(3')-AMP (1 mM) and NaCl (>1 M) at pH 4.5. Pure enzyme with a specific activity of 60 was obtained in only four steps and the yield was about 10 times higher than that of the previous purification method. This enzyme preparation was found to be heterogeneous in molecular weight and was separated into two fractions on Sephadex G-75 gel filtration.

The smaller enzyme with a molecular weight of 36,000 was identical with RNase T_2. The larger one, with an apparent molecular weight of 80,000 is tentatively designated as high-molecular-weight RNase T_2 (RNase T_2-L). RNase T_2-L was still heterogeneous and wad separated into five fractions, RNases T_2-L 1-5, by repeated Sephadex G-150 gel filtration. The amino acid and carbohydrate analyses revealed that each of these fractions has a protein moiety in common with RNase T_2 and the heterogeneities were due to the carbohydrate content, primarily galactose content.

A base-specific RNase Lu (EC 3.1.27.5) was isolated and purified from *Aspergillus niger* in a yield of 23% by means of acetone precipitation, column chromatography on Duolite A-2 DEAE-cellulose, and 2'(3')-aminohexyl-5'-ATP-Sepharose 4B [494] (Fig. 17).

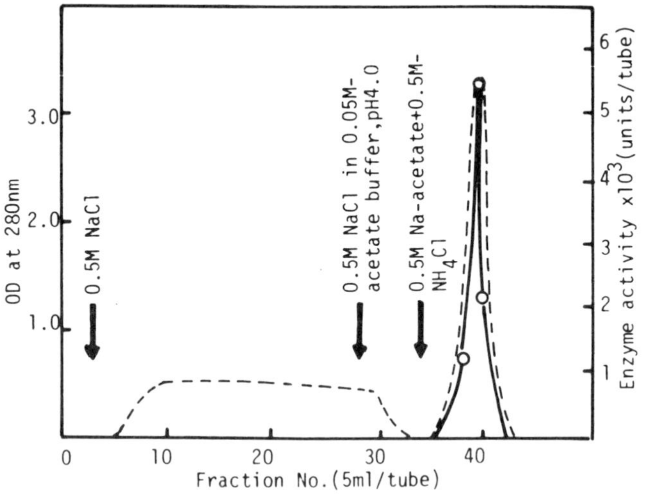

Figure 17 Elution pattern of the enzyme on 2'(3')-aminohexyl-5'-ATP-Sepharose 4B column chromatography. —○—, enzyme activity; ---, protein. (From Ref. 494.)

The enzyme was shown to be homogeneous by disk-polyacrylamide gel electrophoresis. The enzyme released only 3'-UMP from yeast RNA or poly(U) and, in addition, small amounts of 3'-CMP from poly(C). The effects of endonuclease A 236, exonuclease A 5, and pancreatic DNAse on immobilized denaturated DNA fragments have been studied [95].

Lysates of barley chloroplasts release more radioactivity into acid-soluble form from UV irradiated and alkylated-depurinated *E. coli* [^{3}H]DNA than from intact DNA. By means of chromatography on depurinated DNA-cellulose and/or UV irradiated DNA-cellulose and by electrophoresis in polyacrylamide gels, four activities on depurinated DNA were separated [368]. One of these contained activity against heavily UV-irradiated (270 J/m^2) native DNA. In addition, two other nucleases specific toward UV-DNA were separated. One of them was active on native and heat-denatured DNA irradiated with 10 J/m^2 UV, whereas the other was predominantly active on native UV-irradiated DNA.

The extract from rat liver chromatin contains two apurinic/apyrimidinic (AP) endodeoxyribonucleases, named 0.2 M and 0.3 M isozymes according to the phosphate concentration necessary

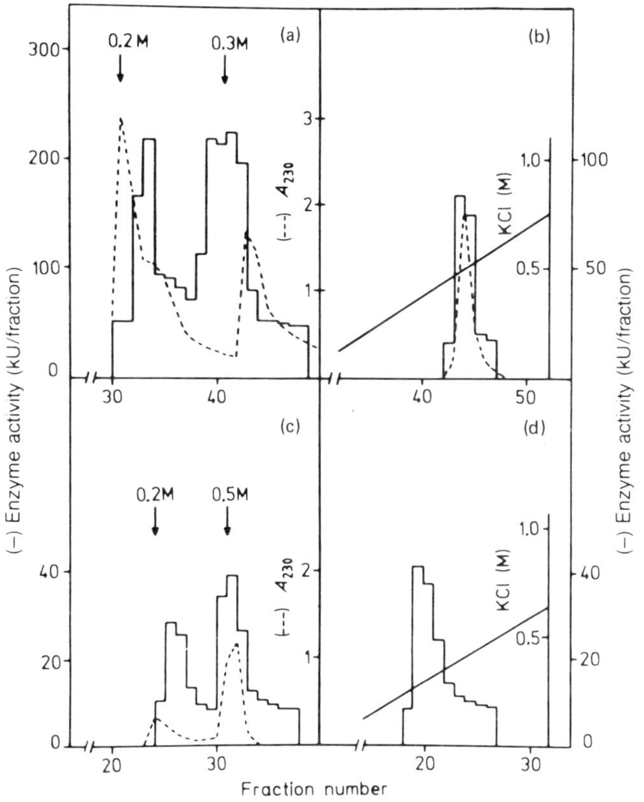

Figure 18 Purification of the 0.3 M isozyme of AP endodeoxyribonuclease. (a) Hydroxylapatite chromatography (1.6 × 15 cm). Elution was carried out with 100 ml of 0.2 M and 100 ml of 0.3 M potassium phosphate, pH 6.8. (b) Fractions 39-43 from the hydroxylapatite chromatography were pooled and dialyzed against buffer B. The solution was poured on a column (1 × 8 cm) of phosphocellulose. After washing with 40 ml of buffer B, the column was eluted with 100 ml of a 0-1 M KCl linear gradient in buffer B. (c) Fractions 44 and 45 from the phosphocellulose chromatography were pooled and dialyzed against 10 mM phosphate, pH 7.30. The solution was poured on a column (1 × 6 cm) of heparin-Sepharose. After a 65-ml washing the column was eluted by steps with 20 ml each of 0.2 M, 0.5 M, and 1.0 M KCl in the same buffer. (d) Fractions 31-33 from the heparin-Sepharose chromatography were pooled and dialyzed against buffer D. The protein

to elute them from an hydroxylapatite column [495] (Fig. 18). The 0.3 M isozyme is the primary and perhaps the only chromatin AP endodeoxyribonuclease in the living cell. This 0.3 M isozyme was purified by successive chromatographies on hydroxylapatite, phosphocellulose, heparin-Sepharose, and alkylated-depurinated DNA-cellulose. The 0.3 M isozyme has no action on intact DNA strands or on alkylated sites; it cuts the phosphodiester bridge, which is the immediate neighbor of the AP site on its 5' side, leaving 3'-hydroxyl and 5'-phosphate ends. It has no associated exonuclease activity. To hydrolyze the phosphoester bond near the AP site, the enzyme makes a close contact with three base residues in the large groove of the DNA molecule. An activity from mouse liver that catalyzes the disappearance of O^6-methylguanine from DNA methylated with methylnitrosourea has been partially purified by ammonium sulfate fractionation and DNA-cellulose chromatography [379] (Fig. 19). DNA-cellulose was prepared from calf thymus DNA [55]. The activity is specific for the repair of O^6-methylguanine lesions and does not affect the removal of 7-methylguanine, 7-methyladenine, or 3-methyladenine.

E. Histone-Specific Acetyltransferase

Three histone-specific acetyltransferases have been characterized in *Artemia* by the criteria of cell compartmentation, chromatographic behavior, substrate specificity, and regulatory properties [503]. Acetyltransferase I is a chromatin-bound enzyme with affinity for DNA-cellulose. The enzyme can acetylate histones H1, H3, and H4, but the acetylation of H1 is markedly inhibited in the presence of H4. Acetyltransferase II and III are cytoplasmic and were resolved by phosphate elution from hydroxylapatite. The isoenzyme II is highly specific for histone H4. The acetyltransferase III is active with the three histone fractions, but its specificity is modulated through the cooperation of H4 (as inhibitor of the acetylation of H1) and H1 (as activator of H4 acetylation). Spermine was confirmed as a specific activator of the acetyltransferase I, with subsaturating concentrations of H3 as substrate.

Figure 18 (Continued) solution was poured on a column (1 × 10 cm) of alkylated-depurinated DNA-cellulose. After washing with 25 ml of buffer D, the column was eluted with 100 ml of a 0-1 M KCl linear gradient in buffer D. (From Ref. 495.)

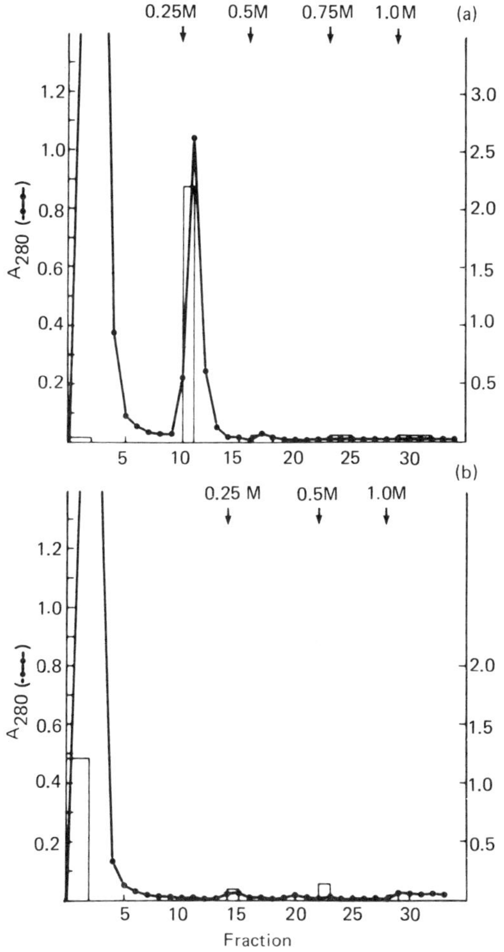

Figure 19 The ammonium sulfate fraction was chromatographed on (a) a DNA-cellulose column or (b) a cellulose column. (From Ref. 379.)

II. ISOLATION OF PROTEINS INVOLVED IN THE TRANSCRIPTION, REPLICATION, AND REGULATION OF NUCLEIC ACIDS

A. DNA-Binding Proteins

The bacteriophage T4-induced "gene 32 protein" appears to be required both for DNA replication and for genetic recombination in T4 bacteriophage-infected *E. coli*. DNA-cellulose was used for the isolation of this protein [53].

The bacteriophage T4 helix-destabilizing protein, the product of gene 32, has been immobilized on an agarose matrix and used for chromatography of lysates of T4-infected *E. coli* cells [514]. At least 10 T4-encoded early proteins and three or four host proteins are specifically retained by this gene 32 protein column. Nine of the T4 proteins have been identified as being involved in either DNA replication or genetic recombination. Notably, the T4 DNA polymerase (gene 43 protein) and two major proteins in the recombination pathway (the products of genes uvsX and uvsY) are specifically bound. On a preparative scale, the column is useful for purification of the bound proteins.

A low-molecular-weight (7000), heat-stable protein (HU) that stimulates transcription of bacteriophage λ DNA by *E. coli* RNA polymerase was purified from *E. coli* extracts using DNA-cellulose [369]. The DNA-cellulose was prepared by UV irradiation [370]. HU binds to native DNA, resulting in an apparent thickening of the DNA chains as revealed by electron microscopy. Contrary to DNA unwinding proteins, it causes no destabilization of the double helix. HU differs from transcription factors (H_1, D, etc.) and from the low-molecular-weight ω subunit of the RNA polymerase.

The DNA-binding (dnaB) protein of *E. coli* is required for replication of the bacterial chromosome. The enzyme exerts its function during the initiation and elongation process of chromosomal replication. The dnaB protein is also required for conversion of ϕX174 single-stranded DNA to the duplex replication form. The purification of the *E. coli* dnaB protein by chromatography on nucleotides bound to agarose has been described [19]. The dnaB protein, which contains an associated ribonucleoside triphosphatase activity, binds to immobilized ATP, ADP, and UDP, but not to AMP. The type of linkage of ATP to agarose influences the adsorption, elution, and purification of the enzyme. Optimal purification is achieved using ATP bound to agarose via its oxidized ribose moiety. By this means, the dnaB protein can be obtained at least 95% electrophoretically pure after only three purification steps. The enzyme can be eluted from immobilized nucleoside-5'-di- and triphosphates by ATP, ADP, and pyrophosphate, but not by AMP or orthophosphate. ADP and pyrophosphate, as well as the substrate ATP in high concentration, are at the same time inhibitors of the ribonucleoside triphosphatase. The dnaB complementing and ribonucleoside triphosphatase activities could not be separated from each other by template chromatography, supporting the finding of others that they both reside on the same protein complex, namely a dnaB multimer. The results indicate that the dnaB protein binds to immobilized nucleotides by means of its ribonucleoside triphosphatase, and that at least the pyrophosphate moiety is essential for adsorption as well as elution of the enzyme.

A DNA-binding protein has also been isolated from two cyanobacteria (blue-green algae), *Anabaena* sp. and *Aphanocapsa* sp., using DNA-cellulose [371] which was prepared according to the reported method [370]. It has a molecular weight of about 10,000. The proteins from both strains show immunological identity with a similar DNA-binding protein from *E. coli*. Both the *Aphanocapsa* and *E. coli* proteins form compact, rapidly sedimenting complexes with cyanophage or bacteriophage DNA. The similarities between the proteins from cyanobacteria and *E. coli* suggest a degree of evolutionary conservation comparable to that of the histones of eukaryotes.

DNA-cellulose chromatography has been used to isolate single-stranded binding proteins from *Tetrahymena thermophila* [372] (Fig. 20, 21). Three major proteins that bind to denatured DNA-cellulose were obtained. The predominant protein possesses many of the properties of the helix-destablizing proteins isolated from prokaryotic and eukaryotic sources. The protein facilitates denaturation of the synthetic copolymer poly d(A-T)·d(A-T), depressing the T_m by nearly 40°C. It also permits the renaturation of poly d(A-T)·d(A-T) in high salt concentration. Two other binding proteins have molecular weights of 25,000 and 23,000 in sodium dodecyl sulfate-polyacrylamide gel electrophoresis. The protein with a molecular weight of 25,000 is probably the "M protein" previously isolated from *T. thermophila*, which has been shown to stimulate *Tetrahymena* DNA polymerase. These two proteins failed to show helix-destabilizing, DNA-dependent ATPase, or deoxyribonuclease activities. One molecule of each protein monomer binds to 7 to 10 nucleotides.

The rat liver single-stranded DNA-binding proteins S25 and HD25, isolated by differential DNA-cellulose chromatography, were compared to the high-mobility-group proteins HMG1 and

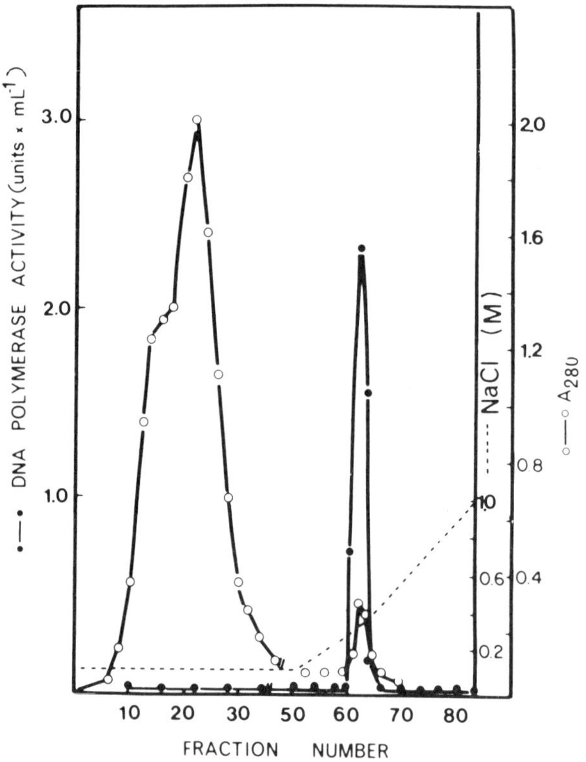

Figure 20 Single-stranded DNA-cellulose (2.5 × 10 cm) chromatography of fraction III (30 ml, 2.0 mg/ml). Elution was with a linear gradient of five column volumes of 0.1-1.0 M NaCl in buffer A at a flow rate of 10 ml/hr. (From Ref. 372.)

HMG2 [496]. Analysis of their amino acid composition, electrophoretic mobility, and tryptic peptide map reveal the identity of the single-stranded DNA-binding protein with HMG1 protein, implying that the rat liver HMG1 protein becomes able both to destabilize a double helix of DNA and to stimulate homologous DNA polymerases only when rat liver cells enter a phase of DNA synthesis, possibly after a specific modification.

The kinetics of Epstein-Barr virus (EBV)-specific protein synthesis has been studied after induction of P3HR-1 cells with n-butyrate [373]. At least four of the polypeptides detected bound

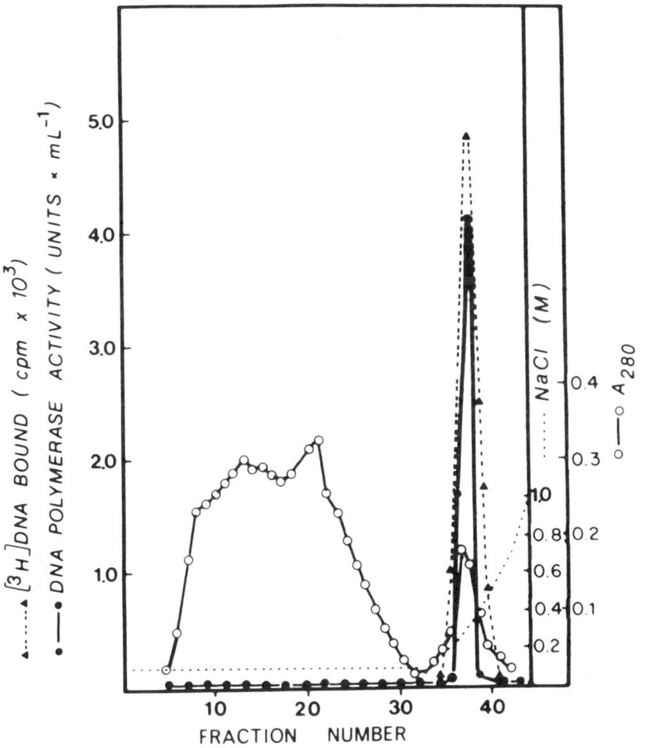

Figure 21 Single-stranded DNA-cellulose chromatography of fraction IV which was chromatographed at the same conditions as those described for Fig. 20. The column was washed (fractions 1-30) and eluted with a 0.1-1.0 M NaCl linear gradient (fractions 31-45). The eluting fractions were assayed for DNA polymerase activity (●) and DNA-binding activity (▲). (From Ref. 372.)

to DNA-cellulose columns. The DNA-binding proteins isolated by DNA-cellulose chromatography represent a subclass of the virus-specific early proteins that can be immunoprecipitated from the total cell extract.

A method for the isolation of sequence-specific DNA-binding proteins based on simultaneous competitive binding to homologous or heterologous DNA has been described [374]. Since both types of DNA are bound to Sephadex beads, the method may be used for the isolation of proteins which bind to specific DNA sequences.

Fractionation of nuclear proteins from Novikoff hepatoma ascites cells by this technique yields two protein fractions which show much greater apparent specificity of binding to homologous DNA, in a nitrocellulose filter binding assay, than unfractionated nucleolar proteins.

A technique for a rapid and efficient isolation and purification of proteins binding to defined DNA sequences has been described [375]. Cloned double-stranded DNA was covalently coupled to m-aminobenzyloxymethyl cellulose to purify proteins which recognize and bind to specific sequences on the DNA. The purification of two DNA-binding proteins from *D. melanogaster* is demonstrated using the respective cloned DNA sequences.

A DNA-binding protein (DB-2) was isolated from unfertilized *Drosophila* eggs [376] using DNA-cellulose [377]. In competition assays with DNA from other species, DB-2 preferentially binds to *Drosophila* DNA. This binding protein can also be isolated from pupal nuclei and comprises only a small fraction (<0.01%) of the total nonhistone chromosomal proteins. DB-2 was used as a probe to screen the gene bank established by inserting random, sheared fragments of *Drosophila* DNA into bacterial plasmids. In plasmid A17, binding of DB-2 protects two short DNA segments of about 13 and 30 bp from digestion by DNase I. These sequences were mapped to a defined restriction fragment. Further evidence for the binding specificity was obtained by use of the electron microscope.

The passing of chromatin fragments of rat liver nuclei through DNA-agarose columns results in the removal of all histones of the H1 class and almost all (>95%) nonhistone proteins from the chromatin and thus leads to the separation of DNA molecules containing nucleosomal histones only [378]. Elution of the proteins bound to DNA-agarose by salt gradients leads to a fractionation of chromosomal proteins, indicating that they bind with different affinities to single-stranded DNA. This procedure is suitable for the isolation of histone III-depleted chromatin and the fractionation of nonhistone proteins.

B. RNA-Binding Proteins of the Cytoplasm

The Qβ host factor, a heat-stable protein necessary in concert with Qβ replicase for phage Qβ RNA replication in vitro, has been localized in *E. coli* and found to be associated primarily with ribosomes [380]. It can be purified by chromatography on poly(A)-cellulose. Ribosomal protein S1 (subunit I of Qβ replicase) also binds to the poly(A)-cellulose column and elutes before the host factor. The Qβ host factor is a potent inhibitor of an in vitro poly(A)-directed polylysine protein-synthesizing system, but has less effect on the in vitro translation of poly(U), R17 RNA, late T7 mRNA, or endogenous *E. coli* mRNA. The amino acid composition and NH_2-terminal sequence rule out the host factor as one of the known 30S or 50S *E. coli* ribosomal proteins. The finding that the Qβ host factor is associated with ribosomes in vivo completes the demonstration that all of the host-supplied proteins required for phage Qβ RNA replication in vitro are either associated with ribosomes or are involved in the protein-synthetic machinery of the cell.

Eukaryotic mRNAs exist in the cytoplasm as protein-bound complexes (mRNP) in polysomes (PmRNP) and in several instances as free nonpolysomal (CmRNP) particles [381] (Fig. 22). The organization and biological function of the proteins in mRNP are not well understood, and the published reports on their variations and diversity remain a matter of considerable dispute. PmRNP derived from a large number of cells contains two major proteins, P52 and P78, and a number of additional polypeptides, usually 2-13. The P78 protein and a protein of similar size have been reported to be bound specifically to the 3'-poly(A) tracts of PmRNP and heterogeneous nuclear RNP. Several conflicting reports have indicated that the sodium dodecyl sulfate-gel electrophoretic patterns of the protein components of PmRNP and CmRNP are either identical or different. This variability may be partially due to differences in the methods of isolation and also to possible contamination by non-mRNP components.

The lighter proteins were reported to have properties similar to those of one of the subunits of the initiation factor IFMP or eIF-2. However, it has been shown that the lighter protein and the β subunit of eIF-2 are not identical.

Approximately 2% of the proteins solubilized from rat liver microsomes or rapidly sedimenting endoplasmic reticulum (RS-ER) adsorbed to poly(A)-Sepharose [382]. The adsorption appeared to be selective for a few proteins, and proteins of different apparent molecular weights adsorbed from RS-ER and the microsomes. The proteins from RS-ER with affinity for poly(A) were coupled to Sepharose and used for the purification of mRNA from rabbit mammary glands. A portion of the RNA that did not adsorb to poly(U)-Sepharose adsorbed to protein-Sepharose and was active in a cell-free protein synthesis system.

The cytoplasm of HeLa cells contains a series of proteins with affinity for RNA. These proteins were detected by template chromatography of RNA-Sepharose [383]. Several proteins are tightly bound to RNA; there is no preference for mRNAs as the same proteins bind to any free RNA, even poly(A).

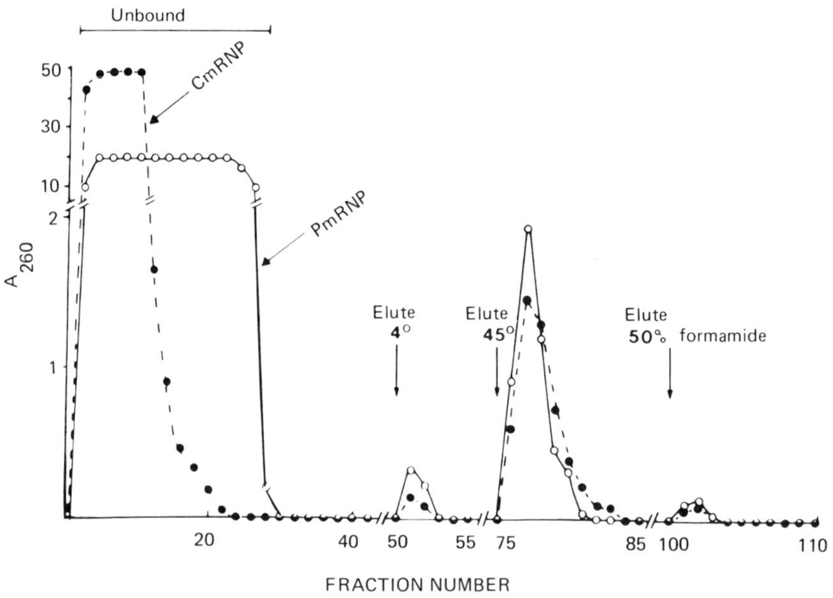

Figure 22 Isolation of free and polysomal mRNP particles by thermal chromatography on oligo(dT)-cellulose of subcellular fractions of chick embryonic muscles. About 500 A_{260} units of EDTA-dissociated 0.5 M KCl-washed free polysomes (○—○), or postpolysomal particles (●—●), used as the sources of polysomal and free mRNP, respectively, dissolved in the binding buffer (10 mM Tris HCl, pH 7.6; 1 mM EDTA; and 0.4 M NaCl) were chromatographed on columns of oligo(dT)-cellulose (0.6 × 3 cm) equilibrated with binding buffer. (From Ref. 381.)

Polysomes of HeLa cells were adsorbed on poly(U)-Sepharose columns [384]. This adsorption is probably due to poly(A) sequences at the 3' terminus of the messenger. Stepwise elution initially removed ribosomal subunits; thereafter, mRNA and a set of proteins. These proteins are identical with the main components of the PmRNP.

The organization of polysomal poly(A)-ribonucleoprotein complex [poly(A)-RNP] was studied. The poly(A)-associated proteins were liberated directly from the poly(A)-RNP complex. Polysomal mRNA was isolated from L5178y mouse lymphoma cells by oligo(dT)-cellulose chromatography [473].

As an initial step toward understanding the role of mRNP complexes in translational regulation during compensatory renal hypertrophy, the characteristics of polysome-associated mRNP isolated by template chromatography were studied [385]. mRNP can also be isolated using oligo(dT)-cellulose from normal and adenovirus-infected KB cells [386].

Mouse globin mRNP has been isolated from reticulocyte polysomes by oligo(dT)-cellulose [387] according to a modified procedure [386]. The mRNA and protein moieties of the mRNP are indistinguishable from those isolated by less rapid techniques such as zonal ultracentrifugation.

Polypeptides associated with the rabbit reticulocyte poly(A) protein complex, isolated by oligo(dT)-cellulose chromatography, were analyzed by sodium dodecyl sulfate-polyacrylamide gel electrophoresis [388]. The complex derived from EDTA-released total polysomal mRNP and 20S globin mRNP by RNase treatment contained four polypeptides.

Nonpolysomal poly(A) mRNP of *Artemia salina* has been isolated by thermal chromatography on oligo(dT)-cellulose [389]. The translatability of the mRNP in a cell-free system depended on the conditions of isolation. The protein composition of the free mRNP is independent of the conditions used in oligo(dT)-cellulose chromatography. A specific set of proteins is associated with different RNP, although some proteins are present on multiple particles. Approximately the same proteins were present on free mRNP and mRNP isolated from nonpolysomal mRNP-ribosome complexes.

mRNP complexes have been isolated from EDTA-dissociated polysomes of *Neuraspora crassa* [390]. Approximately 15% of the [^{3}H] uridine incorporated into polysomal rRNA during a 15-min pulse was eluted from oligo(dT)-cellulose as an mRNP complex.

A study has been made [399] of the developmental changes that occur in the RNA and protein moieties of mRNA-protein particles isolated from newborn and adult rat forebrain free polyribosomes. mRNA-protein particles were isolated by oligo(dT)-cellulose chromatography from salt-washed polyribosomes. The pooled mRNA-protein particles from the E_1 and E_2 fractions after deproteinization with proteinase K sedimented with a mean size of about 18S on a sucrose gradient containing 85% formamide, with little difference between mRNA molecules from newborn and adult. The mean lengths of the poly(A) segments were similar, being about 130 nucleotides long. Distinct changes were found in the protein composition of the mRNA-protein particles. Fractions E_1 and E_2 from the newborn contained two major proteins, with differences in the relative proportions in each fraction. In contrast, adult fractions E_1 and E_2 contained predominantly the larger protein. However, the adult fraction E_2 contained a more heterogeneous population of minor bands of proteins.

Highly purified poly(A) free and polysomal mRNP particles have been isolated by chromatography of subcellular fractions of chick embryonic muscles on oligo(dT)-cellulose and elution with low-salt buffer at 45°C [400-401]. The free and polysomal mRNP represent two distinct classes of macromolecules, the free particles having a more complex nucleoprotein organization than the polysomal particles. Comparison of the protein moieties of three classes of poly(A) cytoplasmic mRNP – those released from nuclei after in vitro transcription and processing (transported mRNP), the free, and the polysomal mRNP – strongly suggests that the majority of the mRNA-associated proteins are exchanged in the cytoplasm during the various functional states of mRNA.

Poly(A) protamine mRNP particles have been isolated from the polysomal and free cytoplasmic subcellular fractions of trout testis cells [391]. The final purification step involved chromatography of pooled gradient fractions of oligo(dT)-cellulose from which intact mRNP could be eluted with distilled water at 40°C. Particles isolated from EDTA-dissociated polysomes were actively translated in vitro, whereas their free cytoplasmic counterparts were not. High-salt washed mRNP particles or the RNA extracted from mRNP preparations, however, directed the synthesis of trout protamines in this system.

Proteins associated with poly(A) RNA in *X. laevis* oocytes were identified in immature ovary homogenates [392]. Four major proteins, $mRNP_{1-4}$, cosediment with poly(A) RNA at 40S-200S, bind to oligo(dT)-cellulose, and are oocyte specific. Two further proteins, $mRNP_{5,6}$, also coisolate with poly(A) RNA. $mRNP_f$ cosediments with poly(A) RNA but fails to bind to oligo(dT)-cellulose. $mRNP_8$ binds to oligo(dT)-cellulose but sediments more slowly than the bulk of poly(A) RNP, at 30S-70S. $mRNP_{1-4}$ are present in immature oocytes but only $mRNP_{3,4}$ are found in full-grown oocytes, while none of them could be detected in *Xenopus* liver or reticulocytes. These proteins are largely cytoplasmic, associated with free mRNP particles, and differ in molecular weight from proteins isolated from polysomal mRNPs in somatic tissues.

Free cytoplasmic poly$(A)^+$ ribonucleoprotein (RNP) was isolated from postpolysomal preparations of maize embryos by chromatography on oligo-(dT)-cellulose [472]. The protein constituents of the RNP were examined by one- and two-dimensional electrophoresis in polyacrylamide gels.

Binding of poly(A) RNP to oligo(dT)-cellulose was investigated as a function of mono- and divalent ion concentration [393]. Eighty to ninety percent binding was obtained either in high (500 mM) or in moderate NaCl concentrations in the presence of 5 mM $MgCl_2$. At 40 mM NaCl and 5 mM $MgCl_2$, poly$(A)^+$ RNP exhibit approximately the same stability as poly$(A)^+$ RNA in binding to oligo(dT)-cellulose, with a T_m of 41°C and 45°C, respectively, indicating that the protein moiety has no effect on RNP binding in these conditions. Differences were observed in the elution of poly$(A)^+$ RNA and poly$(A)^+$ RNP from oligo(dT)-cellulose in buffer without salts. Poly$(A)^+$ RNP was completely removed at 4°C, whereas the T_m of poly$(A)^+$ RNP was only decreased to 34°C.

The interaction of poly(U), poly(A), and poly(C) with small rat liver ribosomal subunits which have protein S6 in different states of phosphorylation was studied by BioGel column chromatography, chromatography on poly(U)-Sepharose 4B, and continuous diafiltration at 4°C [394]. 40S subunits with low phosphorylated protein S6 (isolated from normal liver) and small subunits with highly phosphorylated protein S6 (from galactosamine-, thioacetamide-, dimethylnitrosamine-, puromycin-, and cycloheximide-treated livers) bind initially equal amounts of poly(U), but the dissociation of the radioactive poly(U) occurs much more rapidly and to a greater extent from the low than from the highly phosphorylated type of subunits. From control and galactosamine-40S subunits 62% and 22%, respectively, of originally bound [^{3}H] poly-(U) was removed. The release of initially bound poly(A) from 40S subunits of galactosamine-treated liver was retarded but finally reached the same level as with control liver ribosomal subunits [removal of 40% of bound [^{3}H]poly(A)]. No differences between low and highly phosphorylated subunits were observed with poly(C). If the dissociation reaction was performed at 22°C instead of 4°C, the differences in the release of poly(U) described above disappeared.

RNA-binding proteins isolated from a ribosome-free of amphibian oocytes extract by chromatography on poly(U)-Sepharose possess endogenous protein kinase activity [499]. Incubation of these proteins with [γ-^{32}P]ATP leads to the incorporation of labeled phosphate into six or seven polypeptide chains. High-voltage paper electrophoresis of acid hydrolysates of labeled proteins showed that mainly [^{32}P]phosphoserine is formed in this reaction. The presence of poly(U) in the reaction mixture inhibits phosphorylation considerably.

The ribosomal proteins with affinity for tRNA molecules have been studied [395]. tRNA molecules were covalently coupled to a hydrazinyl-Sepharose 4B. It was found that four ribosomal proteins, S5, S9, L1, and L17, interacted specifically with tRNA molecules. It is suggested that S5, S9, and L2 are directly involved in the binding of tRNA to ribosomes situated at or near the 30S-50S interface and that L17 constitutes part of the seat region of the 50S subunit.

Template chromatography using ribosomal (5S, 16S, 23S) RNA or tRNA covalently coupled to agarose through an adipic acid dihydrazide spacer has been developed to isolate and identify the *E. coli* ribosomal proteins that were specific to these RNAs [113]. Two ribosomal proteins, identified as L18 and L25 by two-dimensional gel electrophoresis, bound tightly to *E. coli* 5S RNA; small amounts of a third protein, L5, also were bound. These proteins did not bind to immobilized tRNA, nor did 30S ribosomal proteins bind to 5S RNA. Denatured 5S RNA bound less L18 and L25 than the native species. Several 30S and 50S ribosomal proteins bind to immobilized tRNA. The major 50S subunit proteins tightly bound to tRNA were identified as L3, L4, L5, L7, and L8/L9. Smaller amounts of proteins L1, L2, L11, L16, and L21 were also observed. Proteins L2 and L16 were retarded by agarose-bound rRNA. Protein S3 was the major 30S subunit protein bound to tRNA. Lesser quantities of proteins S6, S9, S13, and S18 also interacted with tRNA.

Poly(A), poly(C), poly(U), or phage MS2 RNA, immobilized on Sepharose, form a complex with *E. coli* ribosomal proteins [497]. Regardless of their particular nucleotide composition, all four polynucleotides bind an invariable set of proteins consisting of S1, S3, S4, S5, S13, L2, and L17. These polynucleotide protein complexes bind tRNA. Furthermore, it was possible to show that the poly(A)·protein and poly(U)·protein complexes select efficiently their cognate tRNAs, $tRNA^{Lys}$ and $tRNA^{Phe}$, respectively. This important functional property of the polynucleotide·protein complexes suggests that these ribosomal proteins belong in the ribosome to a functional domain responsible for the decoding of mRNA.

Rat liver 5S rRNA was oxidized with periodate and coupled by its 3' terminus to Sepharose 4B through an adipic acid dihydrazide spacer. The ribosomal proteins that bind to that nucleic acid were isolated and identified by electrophoresis in polyacrylamide gels [396]. The eukaryotic 5S rRNA binding proteins were L6 and L19: small amounts of L7, L23', L27/L27', L35', and L39 were also bound to the column, but whether they associate directly and specifically with 5S rRNA is not known. The proteins of the 40S subunit of rat liver ribosomes did not bind to the 5S rRNA column, nor did the proteins of either the large or small subparticle of *E. coli* ribosomes.

The eukaryotic and prokaryotic ribosomal proteins that bind to *E. coli* 5S rRNA were identified [397] (Fig. 23). The *E. coli* ribosomal proteins that associated with the nucleic acid were L5, L18, and L25, confirming earlier findings using the same and different procedures. The rat liver ribosomal proteins that associated with *E. coli* 5S rRNA were L6, L7, L19, L35a, and S9; several of those proteins also bind to rat liver 5S rRNA (L6, L19) and to 5.8S rRNA (L6, L19, and S9).

Wheat germ ribosomal proteins that bind to poly(U), tRNA, and poly(A)-mRNA for leghemoglobin have been identified [498]. The ribosomal proteins that associate with polynucleotides immobilized on Sepharose or on cellulose were identified by polyacrylamide gel electrophoresis. Nine 40S and nine 60S ribosomal proteins were bound to poly(U)-Sepharose. tRNA-Sepharose 4B interacted with five proteins of 40S subunit and with seven proteins of 60S subunit. Six proteins of 60S subunit were bound to poly-(A) mRNA for leghemoglobin. The ribosomal proteins from the small subparticle were not determined because of irreversible binding to the column.

A procedure is described whereby proteins can be covalently and specifically cross-linked to RNA by irradiating isolated ribonucleoprotein [471] cell fractions, or cells with 254-nm UV light. Poly(A)-mRNA-protein complexes and HnRNA-protein complexes are then isolated by chromatography on oligo(dT)-cellulose under conditions that prevent noncovalent binding of proteins, and the proteins cross-linked to RNA are analyzed electrophoretically. A set of at least seven proteins cross-linked to cytoplasmic mRNA has been identified.

C. RNA-Binding Proteins of the Nucleus

The HnRNPs probably play a role not only for the fixation of the HnRNA to the nuclear matrix but also in the processing of HnRNA to mature mRNA. The isolation of HnRNP complex with sedimentation coefficient greater than 80S from rat liver nuclei without the use of RNase inhibitors has been reported [402]. The RNA moiety from these complexes is heterogeneous in size, with a mean sedimentation coefficient of 16S when assayed with polyacrylamide-formamide gel electrophoresis. About 3% of this RNA binds to oligo(dT)-cellulose, the size of the bound fraction being somewhat smaller than the bulk of the HnRNP-RNA. This poly(A)-RNA contains two adenylate tracts, one with about 30 adenylate resi-

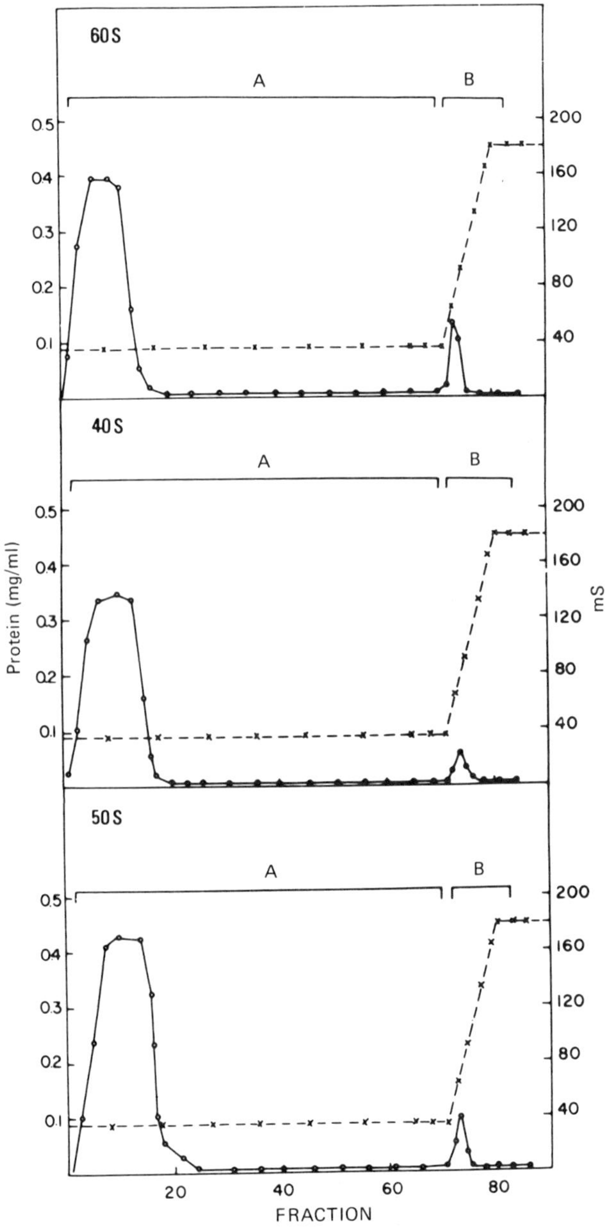
60S
A
B
40S
A
B
50S
A
B
0.5
0.4
0.3
0.2
0.1
200
160
120
80
40
Protein (mg/ml)
mS
20
40
60
80
FRACTION

dues and the second with about 200. Reverse transcription of the poly(A)-RNA and hybridization of the cDNA with their respective templates show two well-defined frequency populations. The same distribution was found whether the hybrids were analyzed with S1 nuclease or hydroxylapatite. Both separated frequency classes hybridize to total genomic DNA. The two populations are also able to hybridize with polysomal poly(A) mRNA, the dilution of both frequencies being very similar in the cytoplasm.

Nuclei isolated from rat liver were incubated under conditions in which RNA continued to be labeled with precursors for long periods. After 1 hr, during which the rate of RNA synthesis was constant, 25-30% of the newly synthesized RNA was recovered in the postnuclear supernatant [403] (Figs. 24 and 25). About 3-5% of this fraction was characterized as poly(A) RNP. The sensitivity of labeling of the RNA synthesized and released from the nuclei to low doses of α-amanitin suggests the presence of polymerase II products in the particles. Comparison of the sizes of proteins in these particles with those of free and polysomal mRNP as well as with HnRNP indicates that the released particles contain a protein of 78,000 daltons, which is also present in the other

---

Figure 23 Chromatography of ribosomal proteins on *E. coli* 5S rRNA-Sepharose. *E. coli* 5S rRNA (4 mg) was oxidized with periodate and coupled to the Sepharose 4B-adipic acid dihydrazide. The 5S rRNA-Sepharose was packed in a column (1 × 5.2 cm), washed with 2.0 M KCl until no substance that absorbed at 260 nm was in the eluate, and then equilibrated with binding buffer (20 mM Tris-HCl, pH 7.4; 0.3 M KCl; 20 mM $MgCl_2$; and 6 mM 2-mercaptoethanol). Ribosomal proteins were applied and the column was washed at 22°C with several volumes of binding buffer at a flow rate of 8 ml/hr. The proteins that did not bind (A) were collected and those that did associate with the nucleic acid (B) were eluted with 30 ml of a linear gradient formed of equal parts of binding buffer and dissociation buffer (20 mM Tris-HCl, pH 7.4; 2 M KCl; 5 mM EDTA; and 6 mM 2-mercaptoethanol). Top: 20 mg of all the proteins of rat liver 60S ribosomal subunits in 40 ml of binding buffer; center: 20 mg of rat liver 40S proteins in 35 ml of binding buffer; bottom: 16 mg of *E. coli* 50S subunit ribosomal proteins in 30 ml of binding buffer. (From Ref. 397.)

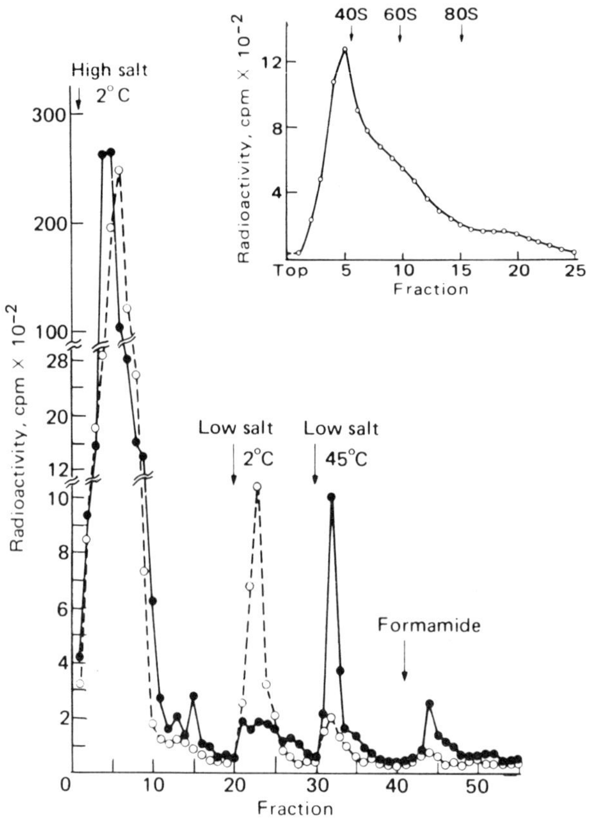

Figure 24 Oligo(dT)-cellulose chromatography of labeled RNA products released in the postnuclear supernatant fraction. RNA was isolated from half of the pooled supernatant, and this RNA and the other half of the supernatant fractions were chromatographed separately. ●—●, Elution profile of the supernatant fraction; ○—○, elution profile of the deproteinized RNA sample. (Inset) Sucrose gradient centrifugation of the labeled products released in the postnuclear supernatant. The arrows indicate the positions of ribosomal subunits and 80S monosomes run as markers in parallel gradients. (From Ref. 403.)

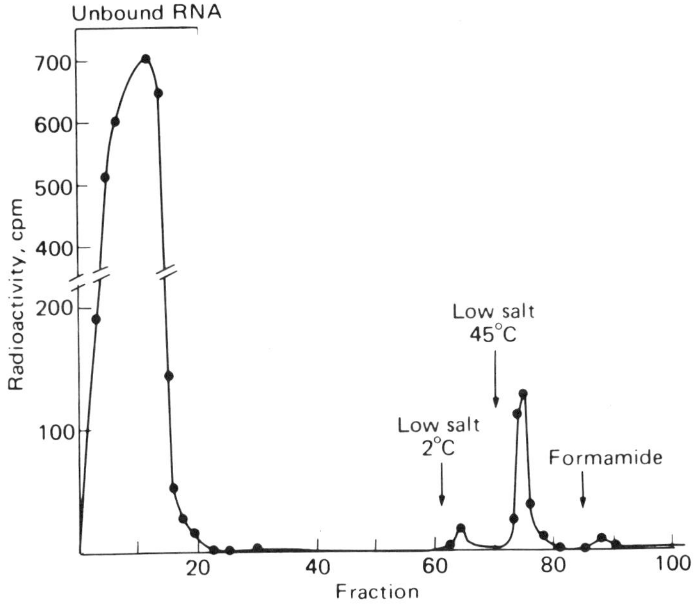

Figure 25 Oligo(dT)-cellulose chromatography of RNase-digested [^{3}H]adenosine-labeled material released in the postnuclear supernatant fraction. (From Ref. 403.)

three classes of ribonucleoproteins, presumably at the 3'-poly(A) segments. In addition, a few other proteins, similar in size to those found in the cytoplasmic RNP, were also present in the released particles.

D. Protein-Synthesis Initiation Factor

The eukaryotic protein-synthesis initiation factor eIF-2 binds GTP and Met-tRNA$_f$ into a ternary initiation complex, eIF-2·GTP·Met-tRNA$_f$. In the presence of 40S ribosomal subunits a quaternary complex is formed. In addition to these basic functions in polypeptide chain initiation, eIF-2 acts as an important site of translational regulation. Studies of the interactions of eIF-2 with GTP, Met-tRNA$_f$, 40S ribosomal subunits, mRNA, and other components under different conditions require substantial amounts of highly purified preparations. Recently, chromatography on mRNA- or double-stranded RNA-cellulose has been introduced on the assumption of a specific interaction of the factor with these immobilized polynucleotides [405,406].

A rapid and efficient procedure has been described for preparing homogeneous polypeptide chain initiation factor eIF-2 from rat liver microsomal salt wash [407] (Fig. 26). The method involves ammonium sulfate fractionation and chromatography on rRNA-cellulose and DEAE-cellulose. The preparation is highly active in initiation complex formation.

Mixed yeast elongator-tRNAs (bulk tRNA lacking $tRNA^{Met}_{m,f}$), pure isoaccepting species of elongator-tRNAs ($tRNA^{Met}_{m}$ and $tRNA^{Phe}$), and purified initiator-tRNA ($tRNA^{Met}_{f}$) were each oxidized with periodate and the 3' terminus was coupled to Sepharose 4B through an adipic acid dihydrazide spacer [398]. The rat liver ribosomal proteins that associated with the tRNAs were isolated and identified (Fig. 27). The rat liver ribosomal proteins

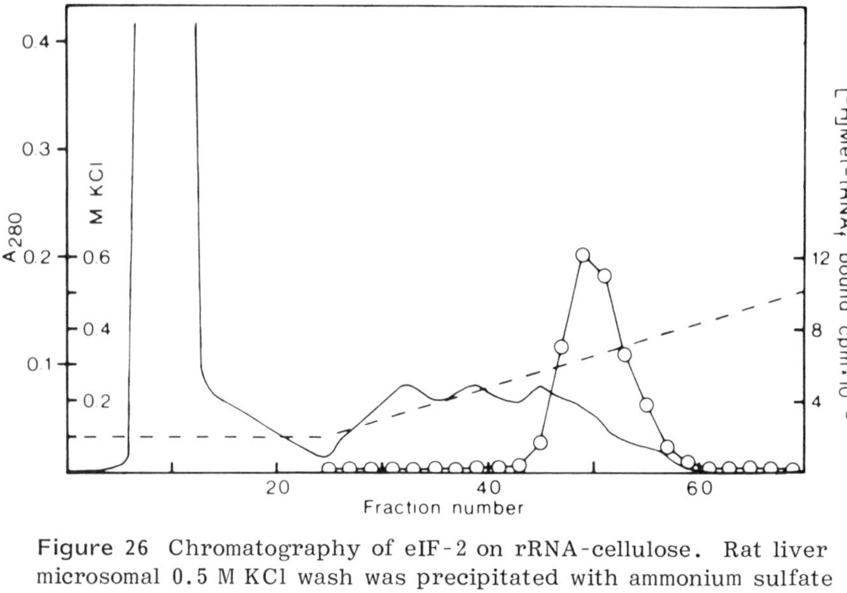

Figure 26 Chromatography of eIF-2 on rRNA-cellulose. Rat liver microsomal 0.5 M KCl wash was precipitated with ammonium sulfate at 40-50% saturation (0°C). The dialyzed material (100 mg of protein) was applied to a 1.5 × 10 cm rRNA-cellulose column equilibrated with buffer A containing 0.1 M KCl. Bound proteins were eluted with a 200 ml 0.1-0.5 M KCl gradient in buffer A. The activities of the fractions collected were measured by ternary initiation complex formation (○—○). —, Absorbance at 280 nm; - - -, concentration of KCl. (From Ref. 407.)

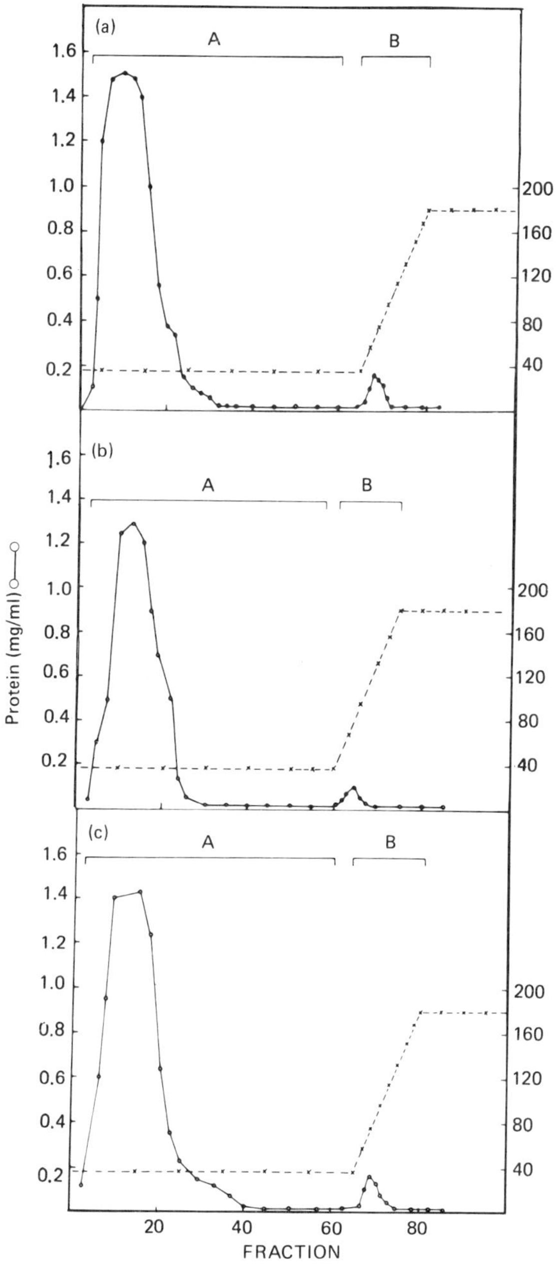

Figure 27 (legend on overleaf)

that were bound to the elongator-tRNA preparations were L6, L35a, and S15; small amounts of a number of other proteins also associated with the nucleic acid. When initiator-tRNA ($tRNA_f^{Met}$) was immobilized on Sepharose, only L6 and L35a were bound; no 40S subunit proteins associated with initiator-tRNA. No *E. coli* proteins formed a complex with either eukaryotic initiator- or elongator-tRNAs.

A poly(U)-binding activity is present in the high-speed supernatant fraction of embryo homogenates from wheat seeds [513]. The factor responsible for such activity was found to have a stimulatory effect on the elongation factor 1 (EF-1). It copurifies with $EF\text{-}1_L$, the lighter form of EF-1, through Sephadex G-200, DEAE-cellulose, hydroxyapatite, and poly(U)-Sepharose 4B column chromatography. The two factors could be separated only through a heating step that destroyed EF-1 activity while leaving most of the poly(U)-binding activity unaltered.

Both myosin mRNA (26S) and globin mRNA (9S) have been bound to activated Sepharose 4B. The affinity of initiation factors derived from native 40S ribosomal subunits from embryonic

---

Figure 27 Chromatography of rat liver ribosomal proteins on tRNA-Sepharose. In (a) 22 mg of 60S ribosomal subunit protein in 45 ml of binding buffer (20 mM Tris-HCl, pH 7.4; 0.3 M KCl; 20 mM $MgCl_2$; and 6 mM 2-mercaptoethanol) was applied to a column (1 × 5 cm) of Sepharose 4B-adipic acid dihydrazide to which 5 mg of elongator-tRNA had been coupled. Chromatography was at 22°C. The ribosomal proteins that did not bind were collected (A) and those that did associate with the nucleic acid (B) were eluted with a linear gradient formed of equal parts of binding buffer and dissociation buffer (20 mM Tris-HCl, pH 7.4; 2 M KCl; 5 mM EDTA; and 6 mM 2-mercaptoethanol). In (b) chromatography was of 18 mg of 40S ribosomal subunit protein in 30 ml of binding buffer on elongator-tRNA-Sepharose just as in (a). In (c), 20 mg of 60S ribosomal subunit protein in 35 ml of binding buffer was applied to a column (1 × 4.5 cm) of Sepharose 4B-adipic acid dihydrazide to which 1 mg of pure yeast initiator-tRNA ($tRNA_f^{Met}$) had been coupled. The conditions for chromatography were the same as in (a). No ribosomal proteins were bound to Sepharose-hydrazine. (From Ref. 398.)

chick muscle for these messengers has been determined [408]. Although both messengers bind the major components of the muscle factor preparation with the same affinity, some differences are noted in the minor components. There is an enrichment of components that bind myosin mRNA with a high affinity when the 15S-18S initiation factor complex is prepared from initiating 40S ribosomal subunits found on myosin-synthesizing polysomes rather than from total cellular factor preparations. The proteins that have a high binding affinity to myosin mRNA also have a discriminating effect when added to a wheat germ system containing myosin and globin mRNA. The components of an impure reticulocyte eukaryotic initiation factor 3, prepared in a similar manner to that for the muscle factor, do not bind myosin mRNA with the same high affinity, and these fractions separated on the myosin mRNA column did not show a discriminatory effect. These results suggest that specific components of muscle 15S-18S initiation factor preparations have a higher binding for myosin mRNA than for globin mRNA, and that these proteins may be those factors reported to be present which discriminate between mRNAs.

10

Enzymic Synthesis or Degradation of Polynucleotides Using Polymer-Bound Primers, Templates, or Substrates

The study of enzymes catalyzing polymerization and depolymerization of polynucleotides is often complicated by the difficulty in differentiating reaction products from starting material. Linkage of polynucleotides to insoluble or soluble polymer supports offers many advantages in studies of these enzymic reactions.

I. ENZYMIC SYNTHESIS OF POLYNUCLEOTIDES USING INSOLUBLE POLYMER-BOUND PRIMERS AND TEMPLATES

DNA polymerase, RNA polymerase, and terminal nucleotidyltransferase can interact with polynucleotides linked to cellulose [409]. The advantages of the polynucleotide cellulose system include the known polarity of primer and product, the ease of quantitative collection and separation of the insoluble template and complementary product, and the capacity to distinguish between reactions occurring free in solution and those involving components fixed to a solid-state matrix. Oligo(dT) in 5'-ester linkage to cellulose matrix serves as primer and template for the *E. coli* DNA polymerase, as a template for RNA polymerase, and as a primer for the calf thymus terminal deoxynucleotidyltransferase. Synthesis by DNA polymerase of the homopolymer pair poly(dA)·poly(dT) takes place when oligo(dT)-cellulose is hydrogen-bonded to an oligo(dA) to form a template pair. The product contains deoxythymidylate covalently linked to the cellulose and deoxyadenylate hydrogen-bonded to the deoxythymidylate. Block copolymers formed by transferase extension of an oligonucleotide cellulose can form hybrids with other synthetic polymers according to base pair specificities. Single-stranded DNA extended with a homopolymer by terminal transferase can hydrogen-bond via the homopolymer region to a complementary derivative of polynucleotide cellulose. Upon replication with DNA polymerase, the product strand is covalently attached to the cellulose.

The complex between a polynucleotide cellulose and a complementary polynucleotide has been investigated as an assay for covalent joining of polynucleotide strands [410]. The starting material is the interrupted homopolymer pair of dI·dC. One strand of poly(dC) is covalently linked to the 3'-terminal hydroxyl group of oligo(dT) bound to cellulose and the other is [^{3}H]-labeled. Alkaline treatment removes the labeled strand from the cellulose unless the interruption has been enzymatically repaired. Labeled strands covalently attached to cellulose are collected by filtration of the alkaline mixture and provide a measure of the number of successful strand ligations.

In the presence of *E. coli* RNA polymerase and all four nucleoside triphosphates, DNA bound to Sephadex G-200 could direct the synthesis of RNA [94]. However, the template capacity of the bound DNA was only one-tenth of that of an equivalent amount of free DNA, in the presence of subsaturating amounts of RNA polymerase. The molecular weight of *E. coli* RNA polymerase is about 500,000, and it seems possible that the presence of the DNA within the matrix of the Sephadex G-200 could reduce the exclusion limit of the Sephadex; therefore, only some of the DNA might be available to bind very large complex proteins. Digestion of the DNA-Sephadex G-200 with pancreatic DNase removed more than 95% of the bound DNA.

In a primer-dependent replication, the reaction stops when the duplex has been completed and further synthesis needs separation of the duplex and a restart in the presence of an excess of the primer. An approach was started with the aim that if a synthetic template (e.g., a single-strand of a gene) could be linked in the proper direction to an insoluble support, then replication and transcription could be achieved at the proper site by providing complementary deoxy- or ribooligonucleotides as primers. After completion of one round of polycondensation, the synthetic polynucleotide could be isolated simply by denaturing the duplex and separating the polymer bound template from the solution. The cycle can be repeated many times. The total process of replication could be much more convenient and very economical with respect to the primer. The following experiment

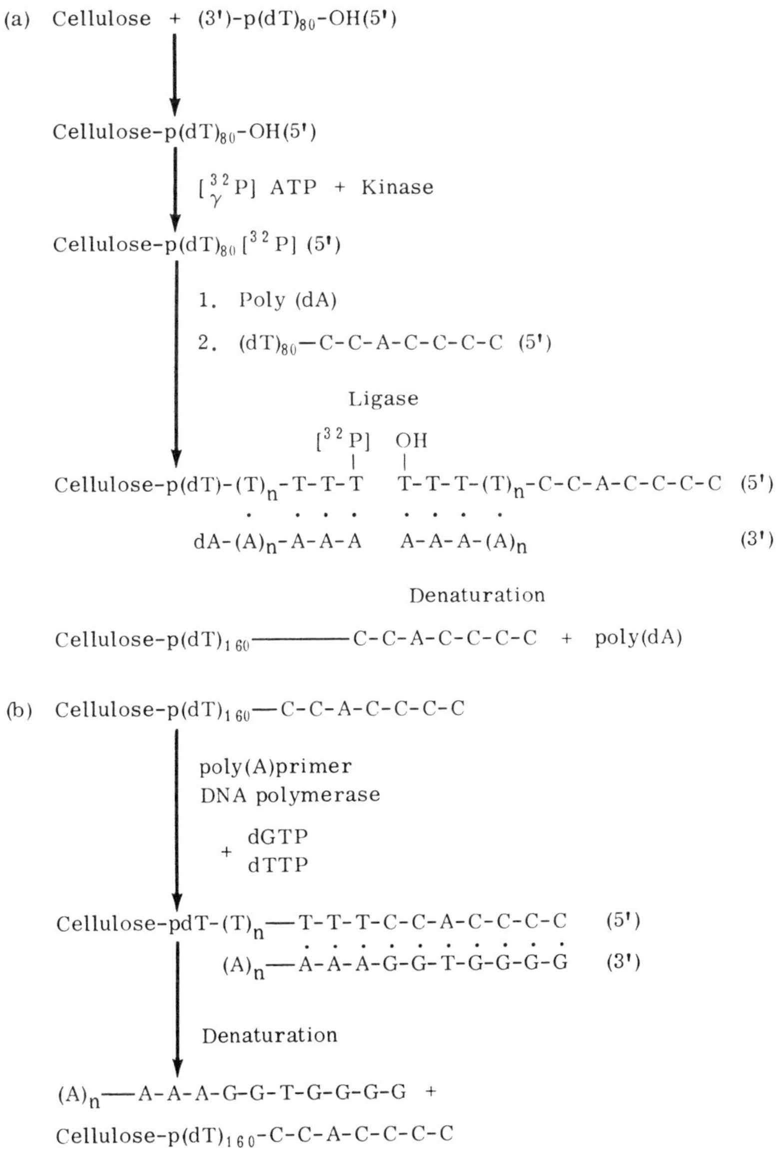

Figure 28 (a) Sequence of reactions used in the linking of d(C-C-C-A-C-C) to cellulose; (b) its replication. (From Ref. 411.)

(Fig. 28) was carried out to investigate this possibility. Poly(dT) ~80 containing 3'-phosphate end groups was immobilized to cellulose. The free 5'-hydroxyl groups of the immobilized poly(dT) ~80 were phosphorylated using [^{32}P]ATP and the polynucleotide kinase. A synthetic deoxyribooligonucleotide segment corresponding to a portion of the *E. coli* tyrosine tRNA gene d(C-C-C-C-A-C-C) was extended at its 3'-OH end by growing a poly(dT) chain using terminal deoxynucleotidyltransferase. The thymidine at the 3' end of this product was covalently linked to the 5'-phosphate end (^{32}P-labeled) of the cellulose-linked poly(dT) by using polynucleotide ligase and poly(dA) as the template. Attention may be drawn to an aspect of the T_4-polynucleotide ligase reaction. When poly(dT) linked to cellulose and carrying a 5'-phosphate group was hybridized with poly(dA) of chain length clearly longer than poly(dT) and the hybrid was treated with the ligase in the presence of ATP, considerable formation of the pyrophosphate between AMP and the 5'-phosphate group of poly(dT) was observed. Thus the enzyme will activate the terminal 5'-phosphate group of the polynucleotide hybridized to a longer complementary chain in the absence of the 3'-OH-carrying acceptor polynucleotide chain. Replication of the specific sequence d(C-C-C-C-A-C-C) in the cellulose-linked polynucleotide above was investigated by using a short $(A)_n$ or poly(dA) as primers suitably hybridized to the poly(dT) region. The product of replication was readily separated from the template after denaturation and was characterized by nearest-neighbor frequency analysis. Another useful application of this approach may be mentioned. Deoxyribopolynucleotides of unknown sequence may be linked to poly(dT) and then immobilized to cellulose. Their replication or precise transcription will be possible simply by the use of oligo(dA) as primers.

The transcription of a mouse globin mRNA by RNA-dependent DNA polymerase can occur using oligo(dT)-cellulose as primer (Fig. 29). The resulting product is probably a single-stranded DNA complement of the mRNA template (or part of it) covalently bound to the cellulose via its 5' terminus through a stretch of oligo(dT). All four deoxyribonucleotide triphosphates, NaCl, the globin mRNA template, and an oligo(dT) primer were required for optimal synthesis of cDNA. By saturating the primer sites using a three-fold excess of mRNA, sufficient concentrations of immobilized cDNA could be synthesized to allow the hybridization reactions to be performed using an excess of globin cDNA. The sequence specificity for hybridization to cDNA-cellulose was determined using 28S rRNA, poly(A), and mouse L-cell RNA. Both

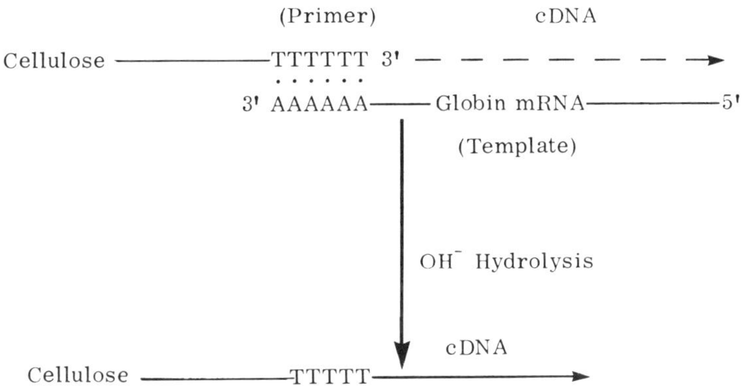

Figure 29 Synthesis of cDNA with oligo(dT)-cellulose as a primer (From Refs. 227 and 412).

analytical and preparative applications of this chromatographic medium were explored. Hybridizations utilizing unfractionated RNA extracts from either mouse reticulocytes or nucleated erythroid cells provided a one-step purification method for globin mRNA sequences. A 60-fold purification of the globin sequences present in reticulocyte polysomal RNA and a 280-fold enrichment of the globin mRNA in nucleated erythroid cells have been found. The globin mRNA isolated by hybridization to cDNA-cellulose retained its biological activity when assayed in a wheat germ cell-free lysate.

A method is described to quantitate labeled globin mRNA by hybridization with excess cDNA which was enzymatically polymerized on oligo(dT)-cellulose [413]. In a large excess of cDNA-cellulose the rate of RNA hybridization was dependent on DNA concentration and not on RNA concentration. Nonhybridized RNA can be digested by RNase and washed from the cDNA which is covalently bound to cellulose. This enables the detection of labeled globin mRNA even when present in a proportion as low as 0.02-0.03% of the total RNA.

The possibility of RNA synthesis by *E. coli* RNA polymerase using denatured DNA templates from mouse liver immobilized on nitrocellulose filters has been shown [564]. The length of the RNA molecules was found to be about 30 nucleotides. Data from alkaline hydrolysis and sedimentation in sucrose density gradient suggest that there is no connection between the DNA primer and the RNA products; therefore, the DNA primer is not necessary for the initiation of RNA synthesis.

Radioactively labeled RNAs were synthesized from cellulose-bound cDNA templates using *E. coli* RNA polymerase [414]. Hybridization of this RNA to excess unlabeled cDNA approached 100%, indicating the complementarity of product and template. The average length of the RNA product, as determined by formamide gels, was approximately 40% of the template length. Hybridization of unlabeled globin RNA produced by this technique to labeled glovin cDNA indicated that the population of RNA sequenes represented at least 80% of the template sequences. Approximately 30% of the RNA product by mass contains poly(A) tails as determined by binding to oligo(dT)-cellulose.

This approach has several interesting possible applications. Having once purified a eukaryotoc mRNA by several laborious steps, solid phase-bound cDNA can be synthesized and the same messenger can be purified repeatedly in a single step. Alternatively, it would provide a useful means for further purifying a partially purified mRNA. For example, if the original mRNA preparation was only partially pure but the mRNA in question represented the predominant sequence, the cDNA-cellulose column made on this template would anneal this sequence at a faster rate than any of the contaminating sequences. This same approach could be used in purification of a specific sequence from total genomic DNA. Similarly, it should be possible to prepare cDNA-cellulose with RNA-virus RNA as a template and to use it to isolate virus-specific mRNA from an infected cell.

cDNA-oligo(dT) celluloses were prepared from cDNA copies of polysomal mRNA of Novikoff hepatoma, normal rat liver, and regenerating rat liver [415]. cDNA synthesis with reverse transcriptase was about 46% with respect to input mRNA with oligo(dT)-cellulose primer. The cDNAs of normal liver, regenerating liver, and Novikoff hepatomas were used as matrices for hybridization of different mRNA species. Under the conditions used, degradation of mRNA was not detected. After normalization for homologous hybridization efficiency, 53% and 65% of the Novikoff hepatoma mRNA bound to normal liver and regenerating liver cDNAs, respectively. Under these conditions an average of 82% of mRNA of normal liver bound to regenerating liver cDNA, and 92% of regenerating liver mRNA bound to normal liver cDNA. Analysis of the proteins synthesized in the wheat germ system indicated that the cDNA columns selectively adsorbed some mRNA species.

Ovalbumin mRNA was purified from hen oviduct polysomal RNA by poly(U)-Sepharose chromatography. The ovalbumin mRNA comprised about 60-65% of this total messenger preparation. This mRNA was used as a template for RNA-dependent DNA polymerase using oligo(dT)-cellulose as primer [453].

DNA complementary to ovalbumin mRNA and covalently bound to cellulose has been synthesized. High concentrations of actinomycin D (200 μg/ml) were required to produce 97% inhibition of double-stranded DNA synthesis, but mRNA transcription was only slightly inhibited (14%) [307]. The conditions used for binding of mRNA to cDNA-cellulose permitted complete hybridization of ovalbumin mRNA in 10 min, while stable poly(A)·(dT) hybrids failed to form. When a batchwise procedure of hybridization and elutions was used, the total recovery of albumin mRNA activity applied to the cDNA-cellulose was >98%, indicating little if any degradation of mRNA. Ovalbumin mRNA activity eluted in each chromatographic run was 50 to 70% of that used originally for the synthesis of the cDNA-cellulose. When the total polysomal RNA was subjected to chromatography, the bound fraction consisted of ovalbumin mRNA, rRNA, and material behaving like fragments of ovalbumin mRNA. Applying this fraction to cDNA-cellulose a second time eliminated the rRNA but not the presumptive fragments. Ovalbumin mRNA purified either once or twice was enriched between 43- and 56-fold over polysomal RNA in translational activity.

A procedure for selectively enriching a mRNA population for inducible sequences has been developed. Other than the induced mRNA species, the population of mRNA in control cells is approximately the same as the mRNA population in induced cells [416]. Cytoplasmic mRNA from control cells is bound to oligo(dT)-cellulose and used as a template for reverse transcriptase, the oligo(dT) serving as a primer. After removing the template mRNAs, the cDNA-cellulose column is used to hybridize a population of mRNAs from induced cells. The nonhybridized poly(A)$^+$ RNAs are greatly enriched in the inducible sequences. This technique of hybridization subtraction chromatography has been used to select a mRNA population enriched for the mRNAs inducible by ecdysterone in Schneider's line 2 *Drosophila* cells. This population of RNAs was used to screen a recombinant library. Preliminary results indicate that about 10% of the RNA in the probe population represents ecdysterone-inducible sequences. This method can be used to select induced mRNAs regardless of the way in which the induction is brought about.

Thymosin α_1 precursor [^{3}H]cDNA-cellulose synthesis was carried out by reverse transcription with RNA-dependent DNA polymerase using oligo(dT) bound to cellulose as a primer [480]. Unlabeled thymosin α_1 precursor cDNA-cellulose was synthesized in a preparative scale to be used in template chromatography. Poly(A)-mRNA from calf thymus was subjected to cDNA-cellulose chromatography and an active messenger obtained in a yield of 3% respect to the total thymus poly(A)-mRNA chromatographed. Analysis of the translation products of the purified mRNA has revealed it as the thymosin α_1 precursor messenger.

For the synthesis of oligonucleotides using polynucleotide nucleotidyltransferase from *M. luteus*, U-U as primer had to be linked to hydroxypropylated Sephadex G-50 [44]. Two μmoles (5.5%) of polymer-bound U-U has been elongated to polymer U-U-$(pA)_4$. In another approach, immobilized $p(dT)_3$ and $p(dT)_6$ have been tested for priming using terminal deoxynucleotidyltransferase from calf thymus [44].

II. ENZYMIC SYNTHESIS OF POLYNUCLEOTIDES USING SOLUBLE POLYMER-BOUND PRIMERS AND TEMPLATES

When polymer-bound polynucleotides are used for enzymatic studies, several considerations determine the choice between the soluble and insoluble forms. The major advantage of the solid forms is that rapid separations can be made simply by filtration or by a quick low-speed centrifugation. However, in certain cases, problems may arise because of noncovalent bonding or adsorption of components to the polysaccharide. It has been reported [417] that very small oligonucleotides and mononucleotides linked to cellulose were either sterically inaccessible to the enzyme, polynucleotide kinase, or, alternatively, the enzyme perhaps became adsorbed to or inactivated by the cellulose. A major disadvantage of the solid form is the inability to carry out further physicochemical characterization of a product synthesized on the solid matrix. The soluble forms offered some hope to fulfill this criterion. For this reason, polynucleotides were covalently linked to Ficoll and dextran and these soluble polymer-bound hybrid molecules have been used in enzymatic studies [417]. Since the used soluble polymer matrix has a very low extinction coefficient in the UV region of interest, optical studies are possible. Hydrodynamic measurements, such as the determination of a sedimentation coefficient, can be made in principle, and some rough qualitative conclusions can be drawn from such experiments. The reactions of polynucleotide kinase and DNA polymerase with polynucleotides linked to Ficoll and dextran have been discussed [417].

Soluble poly(vinyl alcohol)-bound oligo(dT) effectively primes the addition of nucleotide residues catalyzed by terminal deoxynucleotidyltransferase, and the extended chains can act as templates in the presence of DNA polymerase I [418]. From this a new route to polynucleotide synthesis on soluble polymer support based on enzymic reactions in homogeneous liquid phase, seems possible, since to the water-soluble Ficoll-linked primers, the poly(vinyl alcohol)-linked oligonucleotides combine the advantages of polymer support synthesis and of synthesis in homogeneous aqueous phase.

III. ENZYMIC DEGRADATION OF IMMOBILIZED POLYNUCLEOTIDES

An assay for determining endoribonuclease activity involving the coupling of high-molecular-weight RNA to Sepharose has been reported [76]. Using carefully defined coupling conditions, the RNA remains susceptible to endonucleolytic attack. The kinetics of RNA hydrolysis can be followed by sedimenting the remaining RNA-Sepharose complex and determining acid-precipitable RNA in the supernatant solution.

A serious limitation of the RNA-Sepharose assay arises in that it is not practical to perform enzyme reactions near the maximum reaction velocity, since it is not yet feasible to prepare sufficiently concentrated RNA-Sepharose to provide readily saturating substrate concentrations. Another problem routinely encountered with RNA-Sepharose substrates is their instability even when stored at 4°C, since radioactive RNA is progressively lost from the substrate. The half-life of these substrates is often between 5 and 10 days. They cannot be frozen, and reducing their lability with chemical preservatives such as sodium azide or chloroform has not been successful. Despite these limitations, RNA-Sepharose substrates have proved useful for assaying fractions obtained from column chromatography of enzyme extracts or for an initial screening or comparison of extracts and fractions from various sources. As many as 50 assays can be run in a few hours, a marked improvement over sucrose gradients.

11

Studies of Peptide-Nucleotide Interactions

A question of present interest in the field of molecular biology is that of the specific recognition of certain nucleotide sequences by proteins. Template chromatography using immobilized nucleotides or peptides provides a very powerful tool for studying the mechanism of peptide-nucleotide interactions.

I. INTERACTIONS OF RIBONUCLEOSIDE-5'-PHOSPHATE WITH IMMOBILIZED AMINO ACIDS

Optically active anion-exchange gels containing cationic L-amino acids attached to poly(vinyl amine) gel particles have been used for the investigation of mononucleotide-amino acid interactions [419]. Working with monomeric species in aqueous media permits the effects of individual factors to be assessed so that more complex interactions between these molecules can be understood and their evolutionary potential evaluated. An amino acid in its cationic form can interact with nucleotides through its ammonium group as well as its side chain, and differences in the ranking of nucleotides should ultimately be attributable to interactions of the nucleotide base and this side chain. The results show that binding between amino acids and nucleotides has a pronounced system of selectivities which seem to depend on both base and side chain. Nucleotides bind better to arginine than to lysine, and GMP binds better than other nucleotides. The highest interactions were obtained with aromatic amino acids. Preliminary results with dinucleotides and trinucleotides show that binding strengths increase as a function of chain length of the oligonucleotides in most but not all cases. Dinucleotide (U-Gp, C-Gp, A-Gp) associations retain varying degrees of hybrid character contributed by the base at the 5' terminus, and trinucleotide binding is a function of both the nucleotide composition and the order of the bases. These data provide information that should prove useful in deducing functional relationships between these two classes of molecules in the complex systems that arose on a primitive earth and in the study of their role in the evolution of the life.

II. INTERACTIONS OF AMINO ACIDS AND PEPTIDES WITH IMMOBILIZED OLIGONUCLEOTIDES

Template chromatography on oligonucleotide-DEAE-cellulose provides a means of directly estimating the interaction of amino acids with nucleotides [420-422]. The retention of amino acids and peptides on oligonucleotide-DEAE-cellulose is dependent on affinities associated with both the polysaccharide matrix and the immobilized polynucleotides. For this reason, all separations must be carried out under identical conditions on oligonucleotide-DEAE-cellulose and on DEAE-cellulose. Template chromatography is performed in 0.5 M NaCl buffer, so that the anion-exchange character of the DEAE-cellulose is practically canceled. For quantitative comparison of the interactions, relative elution volumes V_r (volume of elution per volume of elution of alanine) are used. If a higher retention is observed on oligonucleotide-DEAE-cellulose than on DEAE-cellulose, the difference, ΔV_r, is a measure of the peptide-nucleotide interation, assuming additivity of the effects (Table 14).

Of all the naturally occurring amino acids, only tryptophan and to a lesser extent tyrosine interact significantly with the immobilized oligo(dT) residues under the conditions for base pairing [421]. The homopolymers of tryptophan and tyrosine undergo greater retardation than the monomers, so that the effect is not additive but multiplicative. Thus Tyr-Tyr-Tyr shows an eightfold and Trp-Trp-Trp an approximately 30-fold larger retardation on PVAL $(pdT)_n$-DEAE-cellulose than tyrosine and tryptophan, respectively. The peptide-oligonucleotide interaction decreases considerably when nonaromatic amino acids are present in the peptide. Consequently, naturally occurring peptides and proteins that contain relatively small amounts of tryptophan and

Table 14 Chromatographic Data of Amino Acids and Synthetic Peptides on PVAL-$(pdA)_n$-DEAE-, PVAL$(pdT)_n$-DEAE-, and DEAE-Cellulose Columns[a]

Components of the mobile phase	Elution volume: PVAL-$(pdA)_n$-DEAE-cellulose		DEAE-cellulose		PVAL-$(pdT)_n$-DEAE-cellulose
	ml	V_r	V_r	ΔV_r	ΔV_r
Ala	56	1.00	1.00	0	0
Phe	64	1.14	1.07	0.07	0.05
Tyr	66	1.18	1.20	-0.02	0.05
Trp	79	1.41	1.43	-0.02	0.10
Trp-Ala	72	1.29	1.37	-0.08	0.13
Trp-Tyr	102	1.82	1.89	-0.07	0.11
Tyr-Tyr	70	1.25	1.39	-0.14	0.31
Trp-Trp	181	3.23	2.91	0.32	0.62
Trp-Trp-Ala	128	2.29	2.50	-0.21	0.50
Ala-Trp-Trp	150	2.68	2.55	0.13	0.53
Phe-Gly-Phe-Gly	69	1.23	1.16	0.07	0.08
Phe-Phe-Phe	65	1.16	1.12	0.04	0.05
Tyr-Tyr-Tyr	101	1.80	1.70	0.10	0.41
Trp-Trp-Trp	513	9.16	8.58	0.58	3.25

[a]Column size in all cases is 23 × 2 cm; elution temperature, 0°C.

Source: Ref. 422.

tyrosine compared with the nonaromatic amino acids undergo at the most only slight retardation on oligo(dT)-DEAE-cellulose. Whereas the interaction of oligo(dT) is directed more toward multiple tryptophan groupings, oligo(dA) undergo interaction preferentially with a Trp-Trp sequence within the peptide chain [422]. Sequence isomers Trp-Trp-Ala and Ala-Trp-Trp can be differentiated on olig(dA)-DEAE-cellulose. Oligo(dT)-DEAE-cellulose, however, cannot distinguish between the sequence isomeric peptides.

III. INTERACTIONS OF OLIGONUCLEOTIDES WITH IMMOBILIZED OLIGOPEPTIDES

Do electrostatic interactions between basic residues of proteins and phosphate groups of nucleic acids participate in the selectivity of recognition between these molecules? No specificity has been reported for the binding of tetralysine to DNAs that differed by 34% in A·T content [423]. This result is in contrast to what has been demonstrated by several authors at the polypeptide level. It has been shown that polylysine preferentially and stochiometrically precipitates A·T-rich DNAs [424]. Polyarginine under the same conditions shows a marked preference for G·C-rich DNAs. More recently, similar conclusions were obtained by template chromatography using polylysine and polyarginine immobilized to agarose [425]. At this point the question arises as to the reason for this A·T specificity. It is conceivable that the preference of the A·T-rich DNA is due to the tertiary structure of the lysine-rich histones. It is, however, also possible that the lysine-rich regions of the histones resemble "active centers" which cause the A·T specificity. If the second assumption is correct, oligolysine peptides should interact preferentially with oligonucleotides rich in A and T rather than with corresponding C- and G-rich oligonucleotides. To settle this question the lysine peptide-oligonucleotide interactions have been investigated using template chromatography [426-428]. In this chromatographic procedure cross-linked polyacrylic gels with immobilized lysine and lysine peptides have been used as models for the investigation of lysine peptide-oligonucleotide interactions (Fig. 30). The interaction can be determined directly from the retardation of the mobile oligonucleotides on the immobilized lysine peptides. The advantages of this method over physicochemical methods lie in its simplicity and in the possibility of analyzing multicomponent mixtures under simulated natural conditions. Summarizing the results of this template chromatography, the following conclusions can be this template chromatography, the following conclusions can be drawn. The strength of the lysine peptide-oligonucleotide interactions depends on the number of phosphate groups of the oligonucleotide. The interactions are base specific. The adenylic and thymidylic acids show a higher affinity toward the lysine peptides than do the corresponding cytidylic acids. The interaction increases discontinuously when going from the trimers to the tetramers of thymidylic acids. The observation that lysine-rich histones interact preferentially with DNA rich in A • T can occur on the level of oligonucleotide-oligopeptide and is not necessarily based on the tertiary structure of the lysine-rich histones. A "conformation fit mechanism" is proposed for explaining the specificity of the interactions observed. Selectivity of the lysine peptide-oligonucleotide interaction occurs only if the positively charged groups are arranged in a specific mode. Oppositely charged partners can approach only to the extent permitted by their conformations. The conformation of oligonucleotides depends strongly on the nature of the constituent nucleobases. It can therefore be assumed that the electrostatic forces which principally exist between the ε-ammonium groups of lysine residues and the phosphate groups of the nucleotides can be exploited to differing degrees. In case of an optimal fit of the conformation of both partners, the electrostatic forces can be exploited maximally. Moreover, these forces will be enhanced by additional physical forces always existing between groups in very close contact (e.g., from surface charge densities, hydrophobic and van der Waals forces, etc.). These principally unspecific forces intensify the electrostatic interaction, resulting in the discontinuous increase of the lysine peptide-oligonucleotide interaction in the case of the tetrameric deoxyribothymidylic acid. With the aid of such a conformation fit mechanism, nature can create highly specific interactions by modulation of unspecific forces.

IV. INTERACTIONS OF DNA WITH IMMOBILIZED POLYPEPTIDES

Template chromatography has been described for evaluation of the specificity of basic polypeptides concerning the base pair composition of DNA [425]. The polypeptides were covalently bound to CNBr agarose. Two DNAs, DNA from *Micrococcus lysodeicticus* (25 mol % A • T) and DNA from *Clostridium perfringens* (69 mol % A•T), strongly different in base composition but equal in molecular weight, were loaded and detached by a NaCl gradient. This method has its precursor in the basic polyamino acid-

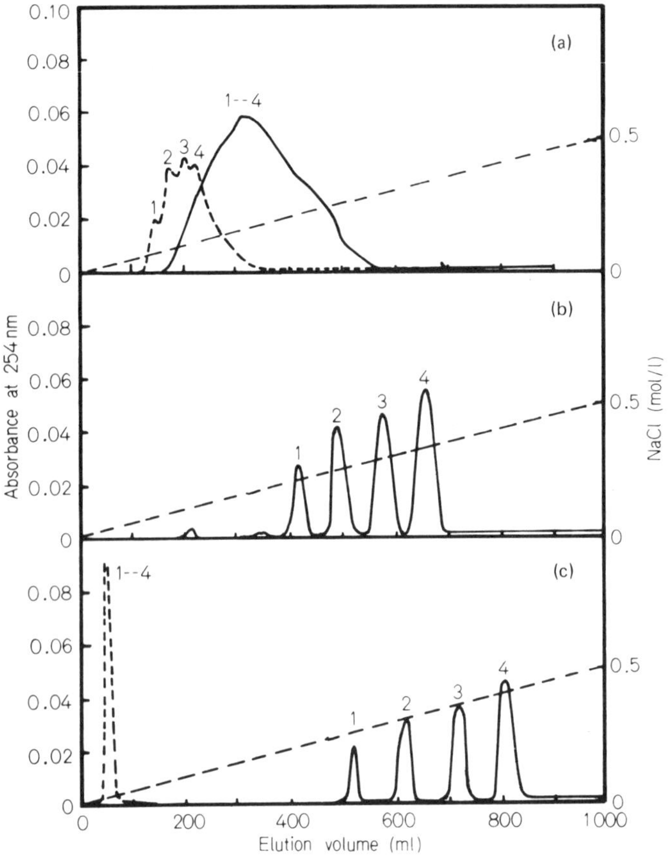

Figure 30 Chromatography of the mixture $p(dT)_{1-4}p$ (a) on gel I (----) and gel II (——), (b) on gel III, (c) on gel IV (----) and gel V (——) using a linear gradient. The mixing vessel contained 0.5 liter of H_2O; the reservoir 0.5 liter of 500 mM NaCl, 25 mM Tris-HCl, pH 7.5. Column size 28 × 2 cm; elution temperature 22°C; flow rate 50 ml/hr. The peak numbers correspond to the individual components of the mixture. (From Ref. 427.)

kieselguhr columns applied for fractionation of different nucleic acids [429-432,545]. In a similar manner, lysine-rich histone fractions were coupled to agarose for studying DNA-histone interactions [433,434]. The differences in the NaCl concentrations

Gel	Structure	Capacity of the wet gel
		μ equiv./g
I	–CO-Lys-OMe	3
II	–CO-$(Lys)_5$ –COOH	21
III	–CO-$(Lys)_5$	15
IV	–CO-$(Lys)_5$-Pro –COOH	24
V	–CO-$(Lys)_5$-Pro	10

Figure 30 (Continued)

between the elution maxima of the two DNAs was taken as a measure for the recognition specificity. The results obtained confirmed the known A • T- and G • C specificity of polylysine and polyarginine, respectively. Neutral residues incorporated into polylysine generally reduce the interaction and also the A • T specificity of their host. This behavior is very pronounced, with three homogeneous fractions of clupeine containing about one-third of neutral aliphatic amino acids within clusters of arginine; the base pair specificity of these arginine copolymers was found to be practically nil.

V. INTERACTIONS OF DNA WITH IMMOBILIZED HISTONE

A simple and rapid method has been described for the purification of supercoiled PM2 DNA by chromatography on columns of H1 histone covalently coupled to agarose [434,563]. The method does not require the use of intercalating agents or ultracentrifugation procedures. Under the conditions most appropriate for purification, elution is carried out in a single step with buffered 0.7 M NaCl. The DNA eluted at the higher salt concentration consists of supercoiled closed circular DNA at about 90% purity, independent of the ratio of supercoiled to nicked circular DNA in the input mixture.

VI. INTERACTIONS OF PEPTIDES WITH IMMOBILIZED DNA

A. Interactions of Polypeptides

Sequential polypeptides $(Lys\text{-}Gly\text{-}Gly)_5$ and $(Lys\text{-}Gly\text{-}Gly)_{10}$ were synthesized as models for the N-terminal region of H4 histone and their interactions with DNA were studied [435]. The ability of $(Lys\text{-}Gly\text{-}Gly)_{10}$ to precipitate DNA was found to be much higher than that of $(Lys\text{-}Gly\text{-}Gly)_5$. For example, at the physiological salt concentration, $(Lys\text{-}Gly\text{-}Gly)_{10}$ precipitates DNA, but $(Lys\text{-}Gly\text{-}Gly)_5$ does not. Both peptides could be acetylated by a partially purified histone acetyltransferase preparation derived from calf thymus. ε-Amino groups of lysyl residues were the sites of the acetylation. DNA-cellulose chromatography of the products indicated that acetylation weakened the interaction of the peptides with DNA.

B. Interactions of Fluorochromes and Proteins

DNA-cellulose was used to determine the quantity of fluorochromes and proteins bound to nucleic acids [436]. The amount of dye bound to the DNA was directly proportional to the A • T base content of the polymer. Of the three acridine derivatives tested, acranil had the highest, quinacrine mustard the lowest, and quinacrine intermediate affinity for the nucleic acids. The affinity of the dye Hoechst 33258 varied in relation to the A • T base content of the polymer. The higher the A • T base content of the DNA, the more poly-L-lysine and the less poly-L-arginine was bound. Binding of polyamino acids or histones to the cellulose-immobilized DNAs partly or wholly prevented the subsequent binding of dyes. Poly-L-lysine was more effective than poly-L-arginine in blocking dye binding to DNA. Complete blocking was obtained with protamine. Of the three histone fractions tested, the lysine-rich fraction was twice as effective in blocking dye binding as the intermediate fraction or the arginine-rich fraction.

C. Interactions of Histones

Template chromatography using DNA-cellulose has been reported for evaluating the strength of binding of histones to DNA [437]. Histones added directly to DNA-cellulose at physiological salt concentrations bind relatively weakly. All histones are eluted together at about 0.5 M NaCl when a salt gradient is applied. However, much tighter binding of the four nucleosomal histones to DNA-cellulose is obtained if gradual histone-DNA reconstitution conditions are used. In this case, the binding of histones H2A, H2B, H3, and H4 to DNA-cellulose closely resembles their binding to native chromatin. The nativeness of the binding is indicated both by the distinctive NaCl elution profile of these histones from DNA-cellulose and by their relative resistance to trypsin digestion when DNA-bound. The binding to DNA-cellulose of histones H2A, H2B, H3, and H4, which have had the first 20 to 30 amino acid residues removed from their NH_2 termini, is indistinguishable from the binding to DNA-cellulose of the same intact histones, as judged by their salt elution profile. Thus, even though the NH_2 termini contain 40 to 50% of the positively charged amino acid residues, a major contribution to the DNA binding comes from the remainder of the histone molecule. Histones can form a "nucleosome-like" complex on single-stranded DNA. These complexes are suggested to resemble biologically important structures which form transiently in chromatin when DNA and RNA polymerases traverse DNA packaged by histones. A unique advantage of the use of the DNA-cellulose method is that it also permits quantitation of histones which are not bound correctly since these appear in a separate peak, eluting at lower salt concentrations. Moreover, since the DNA is attached to a fixed matrix, one can work with these histone•DNA complexes in solvents and salts that would normally precipitate chromatin from solution. However, the certainty that the cellulose matrix does not markably contribute to the histone-DNA interaction has not been checked. Columns containing histone-DNA-cellulose reconstitutes can be used to screen cell extracts for proteins that interact specifically with chromatin. Using such columns, proteins can be found in calf thymus extracts which bind more strongly to histone•DNA complexes than to free DNA [438]. By extension, one could search for proteins that bind only to histone H1, histones H3-H4, or to highly acetylated histones.

D. Interactions of Repressor-Operator

To test the feasibility of site-specific DNA template chromatography, *E. coli* lac repressor was bound to an operator-containing DNA column, and in parallel to a nonoperator DNA column [439]. The elution from nonoperator DNA column was near 250 mmol KCl or NaCl. The elution from operator DNA column, however, was delayed, demonstrating the feasibility of site-specific DNA template chromatography if one provides a sufficiently favorable ratio of specific to nonspecific DNA binding sites.

E. Interactions of Receptors

As a model for the nonspecific interaction of steroid receptors with DNA, the binding of estradiol-receptor complexes of mouse kidney and uterine cytosols to oligo(dT)-cellulose was studied in detail [440]. A limited concentration range of monovalent cationic salts was required for optimal binding, regardless of prior activation of the receptor complexes or the oligomer ligand. Thermal activation of the receptor complexes did not facilitate binding. The reaction was selective for intracellular steroid hormone-receptor complexes, as extracellular proteins binding estradiol with low affinity (bovine serum albumin) or high affinity (mouse α-fetoprotein) were inactive. Both crude and partially purified kidney cytosol receptor complexes bound preferentially to oligo-(dT)- and oligo(dC)-celluloses rather than oligo(dA)-celluloses. These findings suggest that at least part of the nonspecific interaction of estradiol-receptor complexes with native DNA is through a salt-sensitive binding of the complex to the pyrimidine-rich surface of the DNA.

In the C3H mouse mammary adenocarcinoma, estradiol cannot induce the progesterone receptor, and the tumor growth rate is not decreased by ovariectomy [504]. To find an explanation for this estrogen resistance, the authors compared the estrogen receptor (ER) from this tumor to the ER of uterus and of the mammary tumors induced in rats by dimethylbenz[a]anthracene. The binding of ER to DNA-cellulose was analyzed by incubating to equilibrium a constant amount of ER with a variable amount of DNA, the cellulose concentration being kept constant. The authors suggest that the reason for the inefficiency of ER in the C3H tumor may be related to its increased affinity for nonspecific DNA sites.

Androgen receptors from rat kidney and hypothalamus-preoptic area have been demonstrated and partially characterized by their binding properties on DNA-cellulose [441]. For both tissues, the receptors eluted from DNA-cellulose with characteristic patterns, which distinguished the two extracts from one another. In addition, estrogen receptors from kidney were eluted separately from the androgen receptors. The androgen receptors from rat tissues resembled those from mouse tissues in exhibiting limited capacity and high affinity for testosterone and for dihydrotestosterone, in their chromatographic characteristics on DNA-cellulose, and in the distinctive elution patterns between kidney and brain extracts. Tissues from the androgen-resistant rat mutant, "testicular feminization", had reduced levels of androgen receptors in both kidney and hypothalamus-preoptic area. The residual androgen-binding activities in each tissue adhered to DNA-cellulose and eluted with the same patterns as those obtained for receptors of the wild-type rats.

A DNA-cellulose competitive binding assay has been used to measure the extent of displacement of the chicken oviduct progesterone-receptor complex from calf thymus DNA-cellulose by purified cloned fragments of genomic DNA [505]. Several DNA fragments from hormonally responsive genes coding for egg white proteins were found to be efficient competitors for either crude or partially purified receptor complexes when compared with calf thymus DNA. Data obtained with deletion mutations constructed in vitro allowed delineation of a specific region necessary for strong competition, located 250-300 bp upstream from the mRNA start site of the ovalbumin gene. Sequence homologies with this 5'-upstream region were observed in other fragments of the ovalbumin, conalbumin, ovomucoid, and X and Y genes, which were also efficient competitors. Based on a comparison of such sequences of homology, a consensus sequence that may constitute a region binding progesterone-receptor complex has been constructed: $\mathrm{ATC}\genfrac{}{}{0pt}{}{\mathrm{CC}}{\mathrm{TT}}\mathrm{ATT}\genfrac{}{}{0pt}{}{\mathrm{A}}{\mathrm{T}}\mathrm{TCTG}\genfrac{}{}{0pt}{}{\mathrm{G}}{\mathrm{T}}\mathrm{TTGTA}$. The results suggest that specific double-stranded DNA sequences are recognized by the oviduct progesterone-receptor complex in vitro, and are relevant to the question of whether specific DNA sequences are directly involved as genomic binding sites for steroid receptors.

Cytosol glucocorticoid receptors acquired the ability to bind to DNA-cellulose [506]. However, this activation was transient only. It was followed by deactivation whose rate was also dependent on incubation temperature and on the presence of potassium chloride. Deactivation resulted in decreased but nonzero binding to DNA-cellulose. Specific triamcinolone acetonide binding to the receptor protein was not lost under the same conditions. Deactivation commenced before it became apparent, probably together with activation. Preactivated complex underwent deactivation even in conditions that would not allow significant decrease in DNA-cellulose binding without previous incubation at 37°C. In contrast to previous reports, it was found that fast activation/deactivation took place in the presence of 4 mM Ca^{2+}. Molybdate ions slow down both activation and deactivation, but do not prevent activation by heat. Heat-activated/deactivated complex differed in size from both nonactivated and Ca^{2+}-deactivated complexes. This finding suggests that heat and Ca^{2+}-induced deactivation follow different mechanisms.

VII. INVESTIGATION OF ENZYMIC MECHANISM

The mechanism of action of tRNA methylases has been investigated using aminohydroxybutyl cellulose which was linked with each of the individual tRNAs through their oxidized 3' termini [442]. The immobilized tRNAs bind all methylases present in the enzyme extract irrespective of whether this tRNA will be involved in the following step of methylation. These results allow the suggestion that the formation of a methylase-tRNA complex and the catalytic act of methylation are indeed autonomous processes. The first step of interaction between tRNAs and tRNA methylases is rather unspecific and consists of recognition of the whole class of tRNA molecules.

A solid-phase-supported poly(ADP-ribose) polymerase reaction was carried out using purified enzyme bound on DNA-cellulose [443] (Figs. 31A-31C). Essentially all poly(ADP-ribose) synthesized was covalently bound to the enzyme and the enzyme-polymer complex was associated with DNA-cellulose in the presence of Mg^{2+}. The enzyme that is associated with the long polymer could easily be released from DNA-cellulose using a buffer containing EDTA. Mg^{2+} is apparently functioning as a stabilizer of the enzyme-DNA complex in the reaction. Essentially all enzyme molecules were ADP-ribosylated under the Mg^{2+}-dependent reaction condition. The occurrence of the histone modification reaction was also demonstrated in this system. The reaction products with the use of five purified histone classes were all found to be bound more tightly to DNA-cellulose than to the enzyme-bound polymer. The use of an early reaction product of tritium-labeled oligo(ADP-ribosyl) enzyme clearly indicated that no ADP-ribose covalently bound to the enzyme is transferred to histone.

Rats were treated with radioactive dimethylnitrosamine or methyl methanesulfonate and the alkylation of various RNA fractions was determined 4 hr later [444]. It was found that after administration of dimethylnitrosamine, hepatic mRNA (isolated by binding to oligo(dT)-cellulose) and nuclear RNA were alkylated slightly, but significantly less than rRNA. After treatment with methyl methanesulfonate nuclear RNA was alkylated to a greater extent than either m- or rRNA.

The effect of the eukaryotic RNA polymerase inhibitor α-amanitin on RNA synthesis in adult female *Aedes aegypti* has been

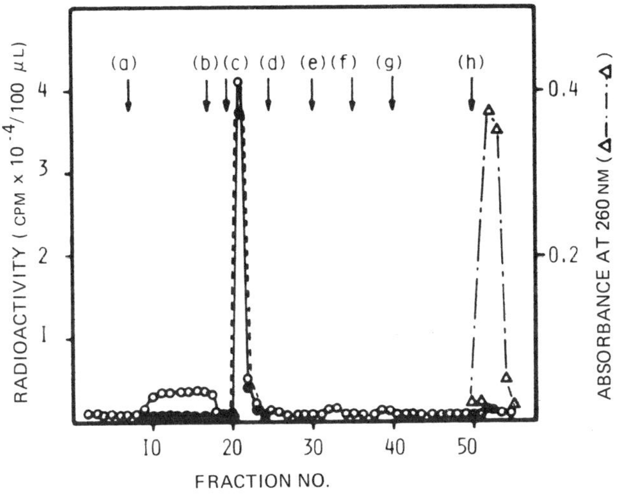

Figure 31A Poly(ADP-ribose) synthesis on a DNA-cellulose column. Enzyme-bound DNA-cellulose containing 180 μg of purified poly(ADP-ribose) polymerase and approximately 1 mg of DNA was packed into a small column (7 × 25 mm). A substrate mixture containing 25 mM Tris-HCl, pH 8.0; 1 mM dithiothreitol; 10 mM $MgCl_2$; and 0.1 mM[^{3}H]NAD^+ (1900 cpm/nmol), as shown by the arrow (a) in the figure, was added to the column at a flow rate of 7 ml/hr at room temperature (23-25°C). After 10 ml of the mixture was passed through, the column was eluted successively with 4 ml of 25 mM Tris-HCl, pH 8.0; 1 mM dithiothreitol; 10 mM $MgCl_2$ (b); 6 ml of standard buffer (c); and 5 ml each of the standard buffer containing 0.2 M (d); 0.4 M (e); 1.0 M (f); and 2.0 M (g); NaCl, successively. After the elution, the column was washed again with the standard buffer and the DNA-cellulose bed was treated with 20 μg or DNase I dissolved in 50 mM Tris-HCL, pH 8.0, and 10 mM $MgCl_2$, in a total volume of 2 ml at 37°C for 30 min, and then the column was eluted with the standard buffer (h). The total (open circles) and acid-insoluble (closed circles) radioactivities of 50 μl of each fraction were counted. (From Ref. 443.)

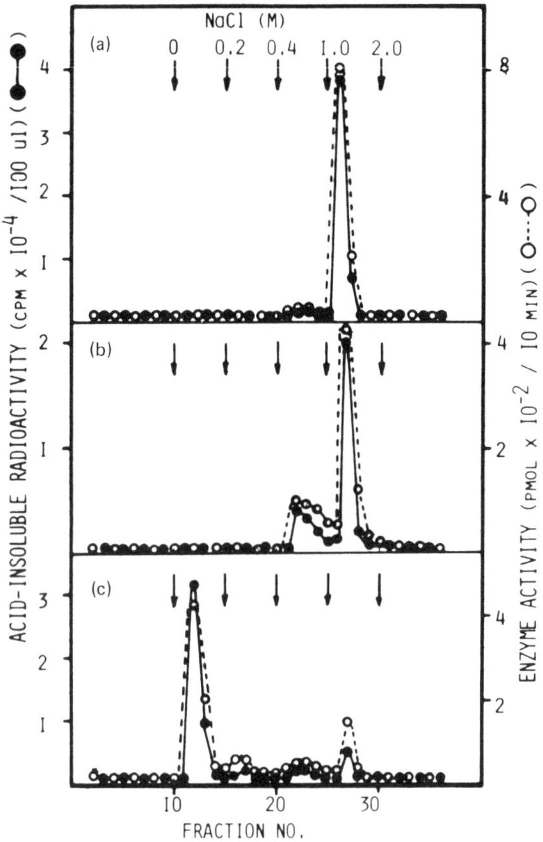

Figure 31B Effect of chain elongation of enzyme-bound ADP-ribose on the binding affinity of enzyme to DNA-cellulose. The enzyme reaction catalyzed by DNA-cellulose-bound poly(ADP-ribose) polymerase was performed under the Mg^{2+}-dependent reaction condition with the use of a 1-ml bed of DNA-cellulose containing approximately 1 mg of DNA, 160 μg of purified poly(ADP-ribose) polymerase in a total volume of the reaction mixture of 4 ml. The concentration and specific radioactivity of substrate [^{3}H]NAD$^+$ were as follows: (a) 0.27 μM (132 × 10^4 cpm/nmol); (b) 10 μM (4.5 × 10^4 cpm/nmol); and (c) 0.1 mM (0.45 × 10^4 cpm/nmol). After the reaction at 25°C with vigorous shaking, the sample was cooled in ice and loaded on a small column (7 mm in diameter). The DNA-cellulose bed was washed with 5 ml of 25 mM Tris-HCl, pH

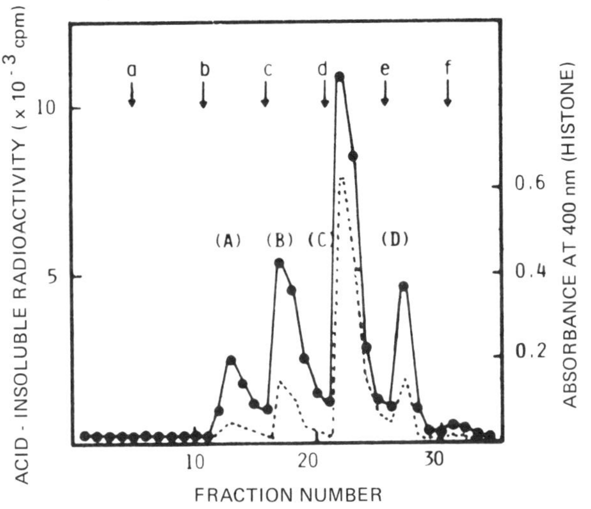

Figure 31C Elution of ADP-ribosylated histone H1 from DNA-cellulose. The enzyme reaction was performed under the histone-dependent reaction condition with 100 μg of enzyme and 1.7 mg of histone H1 bound to nicked DNA-cellulose (1 mg of DNA) in a 10-ml reaction mixture. The concentration of [^{3}H]NAD^+ (6400 cpm/nmol) was 50 μM and the incubation was carried out at 25°C for 20 min. After the incubation, the sample was chilled in ice and loaded on a small column (7 mm in diameter). The column was eluted with 5 ml each of the standard buffer containing 0 (a), 0.1 (b), 0.2 (c), 0.4 (d), 1.0 (e), and 2.0 (f) M NaCl, successively. Acid-insoluble radioactivity (●—●) and histone H1 (---) in the samples were measured. (From Ref. 443.)

Figure 31B (Continued) 8.0; 1 mM dithiothreitol; and 10 mM $MgCl_2$. Elution was performed with 5 ml each of the standard buffer containing 0, 0.2, 0.4, 1.0, and 2.0 M NaCl, as indicated in the figure. (From Ref. 443.)

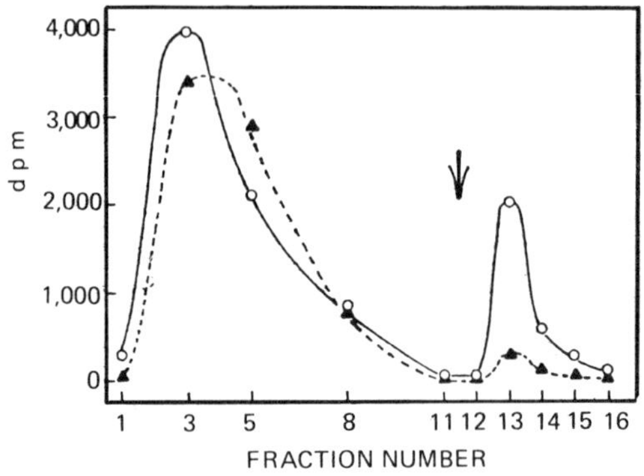

Figure 32 Poly(U)-Sepharose chromatography of microsomal fractions from saline- and α-amanitin-treated females. Females were injected with saline (○—○) or α-amanitin (2 μg/ml, ▲—▲) and 24 hr postinjection they were labeled with [^{3}H]uridine. Microsomal fractions were prepared and were subjected to poly(U)-Sepharose chromatography. Total radioactivity loaded on the column was 21,300 dpm for the saline-treated sample and 18,500 dpm for the α-amanitin-treated sample. (From Ref. 445.)

studied [445] (Fig. 32). Administration of the toxin 24 hr prior to a 2-hr pulse of radioactive precursors inhibited incorporation of the label into total RNA by approximately 40%. Nuclear RNA and microsomal RNA showed 25% and 15% less radioactivity, respectively, than the saline-injected controls. Total soluble 4S RNA was unaffected by α-amanitin, but heterodisperse RNA incorporated 50% less label than controls. Using both nitrocellulose membrane filters and poly(U)-Sepharose chromatography, it was found that the synthesis of poly(A) microsomal RNA was severely inhibited by the toxin. It was concluded that the major long-term effect of α-amanitin in these adults is to inhibit mRNA synthesis and to have little, if any effect on ribosomal or 4S RNA.

References

1. P. V. Sundaram, *Nucleic Acids Res.*, *1*:1587 (1974).
2. R. Barker, I. P. Trayer, and R. L. Hill, *Methods Enzymol.*, *34*:479 (1974).
3. I. P. Trayer and M. A. Winstanley, *J. Biolchem.*, *9*:449 (1978).
4. E. A. Wider De Xifra, S. Mendiara, and A. M. del C. Batlle, *FEBS Lett.*, *27*:275 (1972).
5. K. Mosbach, H. Guilford, R. Ohlsson, and M. Scott, *Biochem. J.*, *127*:625 (1972).
6. C. R. Lowe, M. J. Harvey, D. B. Craven, and P. D. G. Dean, *Biochem. J.*, *133*:499 (1973).
7. E. G. Sander, D. B. McCormick, and L. D. Wreigh, *J. Chromatogr.*, *21*:419 (1966).
8. P. T. Gilham, *Methods Enzymol.*, *21*:191 (1971).
9. P. V. Danenberg, R. J. Langenbach, and C. Heidelberger, *Biochem. Biophys. Res. Commun.*, *49*:1029 (1972).
10. O. Berglund and F. Eckstein, *Eur. J. Biochem.*, *28*:492 (1972).
11. M. A. Michelson, *Biochim. Biophys. Acta*, *91*:1 (1964).
12. D. E. Hoard and D. G. Ott, *Am. Chem. Soc.*, *87*:1785 (1965).
13. R. Barker, K. W. Olsen, J. H. Shaper, and R. L. Hill, *J. Biol. Chem.*, *247*:7135 (1972).
14. I. P. Trayer, H. R. Trayer, D. A. P. Small, and R. C. Bottomley, *Biochem. J.*, *139*:609 (1974).

15. P. J. Hoffman and R. L. Blakley, *Biochemistry, 14*:4804 (1975).
16. J. C. Paulson, W. E. Beranek, and R. L. Hill, *J. Biol. Chem., 252*:2356 (1977).
17. P. O. Larsson and K. Mosbach, *Biotech. Bioeng., 13*:393 (1971).
18. M. Wilchek and R. Lamed, *Methods Enzymol., 34*:475 (1974).
19. E. Lanka, C. Edelbluth, M. Schlicht, and H. Schuster, *J. Biol. Chem., 253*:5847 (1978).
20. M. G. Boulton, A. S. Jones, and R. T. Walker, *Biochim. Biophys. Acta, 246*:197 (1971).
21. R. Lamed, Y. Levin, and M. Wilchek, *Biochim. Biophys. Acta, 304*:231 (1973).
22. P. D. Wagner, *FEBS Lett., 81*:81 (1977).
23. T. Uchida and Y. Shibata, *J. Biochem., 90*:463 (1981).
24. H. Rosemeyer and F. Seela, *Carbohydr. Res., 62*:155 (1978).
25. Y. D. Clonis and C. R. Lowe, *Eur. J. Biochem., 110*:279 (1980).
26. F. Seela and S. Waldek, *Nucleic Acids Res., 2*:2343 (1975).
27. C. R. Lowe, in *An Introduction to Affinity Chromatography: Laboratory Techniques in Biochemistry and Molecular Biology* (T. S. Work and E. Work, Eds.), North-Holland, Amsterdam, 1979.
28. R. E. Scofield, R. P. Werner, and F. Wold, *Anal. Biochem., 77*:152 (1977).
29. S. J. Morris, *J. Chromatogr., 82*:349 (1973).
30. H. Schott, G. Greber, and L. Bucsis, *Makromol. Chem., 136*:303 (1970).
31. H. Schott and G. Greber, *Makromol. Chem., 136*:307 (1970).
32. H. Schott and G. Greber, *Makromol. Chem., 144*:333 (1971).
33. A. S. Jones, D. G. Parsons, and D. G. Roberts, *Eur. Polym. J., 3*:187 (1967).
34. V. N. Shibaev, Yu. Yu. Kusov, N. A. Kalinchuk, and N. K. Kochetkov, *Bioorg. Khim., 3*:120 (1977).
35. P. Gacesa and W. J. D. Whish, *Biochem. J., 175*:349 (1978).
36. P. T. Gilham, *J. Am. Chem. Soc., 84*:1311 (1962).
37. P. T. Gilham, *J. Am. Chem. Soc., 86*:4982 (1964).
38. S. Erhan, L. G. Northrup, and F. R. Leach, *Proc. Natl. Acad. Sci. USA, 53*:646 (1965).
39. P. T. Gilham, *Biochemistry, 7*:2809 (1968).
40. C. R. Astell and M. Smith, *Biochemistry, 11*:4114 (1972).
41. J. C. Smith and M. A. W. Eaton, *Nucleic Acids Res., 1*:1763 (1974).

42. L. Clerici, F. Campagnari, J. F. M. de Rooij, and J. H. van Boom, *Nucleic Acids Res.*, *6*:247 (1979).
43. V. V. Shumyantzeva, N. I. Sokolova, and Z. A. Shabarova, *Nucleic Acids Res.*, *3*:903 (1976).
44. H. Köster, K. Albersmeyer, and D. Skroch, *Hoppe-Seyler's Z. Physiol. Chem.*, *359*:1579 (1978).
45. J. F. M. de Rooij, G. Wille-Hazeleger, A. B. J. Vink, and J. H. van Boom, *Tetrahedron*, *35*:2913 (1979).
46. H. Schott and H. Watzlawick, *Makromol. Chem.*, *182*:825 (1981).
47. H. Schott and H. Watzlawick, *Makromol. Chem.*, *182*:119 (1981).
48. H. Schott, *J. Chromatogr.*, *96*:79 (1974).
49. H. Schott, *J. Chromatogr.*, *115*:461 (1975).
50. P. T. Gilham, in *Immobilized Biochemicals and Affinity Chromatography* (R. B. Dunlap, Ed.), Plenum, New York, 1974, p. 173.
51. H. Potuzak and P. D. G. Dean, *FEBS Lett.*, *88*:161 (1978).
52. W. H. Scouthen, in *Chemical Analysis*, Vol. 59, Wiley, New York, 1981, p. 208.
53. B. M. Alberts, F. J. Amodio, M. Jenkins, E. D. Gutmann, and F. L. Ferris, *Cold Spring Harbor Symp. Quant. Biol.*, *33*:289 (1968).
54. B. M. Alberts, *Fed. Proc. Fed. Am. Soc. Exp. Biol.*, *29*:1154 (1970).
55. B. Alberts and G. Herrick, *Methods Enzymol.*, *21*:198 (1971).
56. R. J. Britten, *Science*, *142*:963 (1963).
57. P. P. Hung, *Science*, *149*:639 (1965).
58. R. Sheldon, C. Jurale, and J. Kates, *Proc. Natl. Acad. Sci. USA*, *69*:417 (1972).
59. G. G. Carmichael, K. Weber, A. Niveleau, and W. Wahba, *J. Biol. Chem.*, *250*:3607 (1975).
60. G. G. Carmichael, *J. Biol. Chem.*, *250*:6160 (1975).
61. R. M. Litman, *J. Biol. Chem.*, *243*:6222 (1968).
62. N. V. Fedoroff and N. D. Zinder, *Proc. Natl. Acad. Sci. USA*, *68*:1838 (1971).
63. I. Smith, H. Smith, and S. Pifko, *Anal. Biochem.*, *48*:27 (1972).
64. E. T. Bolton and B. J. McCarthy, *Proc. Natl. Acad. Sci. USA*, *48*:1390 (1962).
65. A. J. Bendich and E. T. Bolton, *Methods Enzymol.*, *12B*:635 (1968).
66. H. Schaller, C. Nüsslein, F. J. Bonhoeffer, C. Kurz, and I. Nietzschmann, *Eur. J. Biochem.*, *26*:474 (1972).

67. M. Fuke and C. A. Thomas, *J. Mol. Biol.*, *52*:395 (1970).
68. D. Gillespie and S. Spiegelman, *J. Mol. Biol.*, *12*:829 (1965).
69. H. Sano, H.-D. Royer, and R. Sager, *Proc. Natl. Acad. Sci. USA*, *77*:3581 (1980).
70. R. F. Baker, *Anal. Biochem.*, *78*:569 (1977).
71. S. L. Petrović, M. B. Novaković, and J. S. Petrović, *Biopolymers*, *14*:1905 (1975).
72. A. Wada and A. Kishizaki, *Biochem. Biophys. Acta*, *166*:29 (1968).
73. L. F. Cavalieri and E. Carroll, *Proc. Natl. Acad. Sci. USA*, *67*:807 (1970).
74. M. S. Poonian, A. J. Schlabach, and A. Weissbach, *Biochemistry*, *10*:424 (1971).
75. A. Weissbach and M. S. Poonian, *Methods Enzymol.*, *34*:463 (1974).
76. M. V. Berridge and A. I. Aronson, *Anal. Biochem.*, *53*:603 (1973).
77. D. J. Arndt-Jovin, T. M. Jovin, W. Bähr, A.-M. Frischauf, and M. Marquardt, *Eur. J. Biochem.*, *54*:411 (1975).
78. A. F. Wagner, R. L. Bugianesi, and T. Y. Shen, *Biochem. Biophys. Res. Commun.*, *45*:184 (1971).
79. U. Lindberg and T. Persson, *Eur. J. Biochem.*, *31*:246 (1972).
80. S. Okada, Y. Husimi, S. Tanabe, and A. Wada, *Biopolymers*, *14*:33 (1975).
81. P. M. Pitha, *Tex. Rep. Biol. Med.*, *35*:39 (1977).
82. J. Kempf, N. Pfleger, and J. M. Egly, *J. Chromatogr.*, *147*:195 (1978).
83. B. A. Klyashchitsky, V. Kh. Mitina, G. E. Morozevich, and R. I. Yakubovskaya, *J. Chromatogr.*, *210*:67 (1981).
84. B. E. Noyes and G. R. Stark, *Cell*, *5*:301 (1975).
85. H. W. Dickerman, T. J. Ryan, A. I. Bass, and N. K. Chatterjee, *Arch. Biochem. Biophys.*, *186*:218 (1978).
86. G. R. Stark and J. G. Williams, *Nucleic Acids Res.*, *6*:195 (1979).
87. M. L. Goldberg, R. P. Lifton, G. R. Stark, and J. G. Williams, *Methods Enzymol.*, *68*:206 (1979).
88. G. R. Stark, W. Dower, R. T. Schimke, R. E. Brown, and I. M. Kerr, *Nature*, *278*:471 (1979).
89. J. C. Alwine, D. J. Kemp, B. A. Parker, J. Reiser, J. Renart, G. R. Stark, and G. M. Wahl, *Methods Enzymol.*, *68*:220 (1979).
90. B. Seed, *Nucleic Acids Res.*, *10*:1799 (1982).

91. A. J. Macdougall, J. R. Brown, and T. W. Plumbridge, *Biochem. J.*, *191*:855 (1980).
92. E. K. F. Bautz and B. D. Hall, *Proc. Natl. Acad. Sci. USA*, *48*:400 (1962).
93. A. J. Adler and A. Rich, *J. Am. Chem. Soc.*, *84*:3977 (1962).
94. D. Rickwood, *Biochim. Biophys. Acta*, *269*:47 (1972).
95. V. I. Latich, V. I. Varlamov, N. N. Semenova, and S. V. Rogozhin, *Biokhimiya*, *45*:1597 (1980).
96. T. Y. Shih and M. A. Martin, *Biochemistry*, *13*:3411 (1974).
97. W. C. Saxinger, C. Ponnamperuma, and D. Gillespie, *Proc. Natl. Acad. Sci. USA*, *69*:2975 (1972).
98. S. Biagioni, R. Sisto, A. Ferraro, P. Caiafa, and C. Turano, *Anal. Biochem.*, *89*:616 (1978).
99. H.-D. Hunger, H. Grutzmann, and C. Coutelle, *Biochim. Biophys. Acta*, *653*:344 (1981).
100. H. Potuzak and U. Wintersberger, *FEBS Lett.*, *63*:167 (1976).
101. H. Potuzak and P. D. G. Dean, *Nucleic Acids Res.*, *5*:297 (1978).
102. L. G. Moss, J. P. Moore, and L. Chan, *J. Biol. Chem.*, *256*:12655 (1981).
103. G. Zubay, *J. Mol. Biol.*, *4*:347 (1962).
104. J. C. Lee and P. T. Gilham, *J. Am. Chem. Soc.*, *88*:5685 (1966).
105. J. C. Lee, H. L. Weith, and P. T. Gilham, *Biochemistry*, *9*:116 (1970).
106. D. M. Brown and A. P. Read, *J. Chem. Soc.*, 5072 (1965).
107. D. G. Knorre, S. D. Myzina, and L. S. Sandakhchiev, *Izv. Sib. Otd. Akad. Nauk SSSR, Ser. Khim Nauk*, 135 (1964).
108. J. Petre, A. Bollen, P. Nokin, and H. Grosjean, *Biochimie*, *54*:823 (1972).
109. P. Remy, C. Birmelé, and J. P. Ebel, *FEBS Lett.*, *27*:134 (1972).
110. D. L. Robberson and N. Davidson, *Biochemistry*, *11*:533 (1972).
111. K. G. Skryabin, V. P. Varlamov, V. M. Zakhariev, S. V. Rogozhin, and A. A. Bayev, *Bioorg. Khim.*, *2*:1416 (1976).
112. H. R. Burrell and J. Horowitz, *FEBS Lett.*, *49*:306 (1976).
113. H. R. Burrell and J. Horowitz, *Eur. J. Biochem.*, *75*:533 (1977).
114. M. B. Ustav, J. L. Remme, A. J. Lind, and R.L.-E. Villems, *Bioorg. Khim.*, *5*:365 (1979).

115. H. L. Weith, J. L. Wiebers, and P. T. Gilham, *Biochemistry, 9*:4396 (1970).
116. M. Rosenberg, J. L. Wiebers, and P. T. Gilham, *Biochemistry, 11*:3623 (1972).
117. M. Rosenberg and P. T. Gilham, *Biochim. Biophys. Acta, 246*:337 (1971).
118. T. F. McCutchan, P. T. Gilham, and D. Söll, *Nucleic Acids Res., 2*:853 (1975).
119. H. Schott, E. Rudloff, P. Schmidt, R. Roychoudhury, and H. Kössel, *Biochemistry, 12*:932 (1973).
120. E. C. Moore, D. Peterson, L. Y. Yang, C. Y. Yeung, and N. F. Neff, *Biochemistry, 13*:2904 (1974).
121. M. Rosenberg, *Nucleic Acids Res., 1*:653 (1974).
122. S. Hecht, *Tetrahedron, 33*:1671 (1977).
123. R. E. Duncan and P. T. Gilham, *Anal. Biochem., 66*:532 (1975).
124. R. P. Singhal, R. K. Bajaj, C. M. Buess, D. B. Smoll, and V. N. Vakharia, *Anal. Biochem., 109*:1 (1980).
125. B. J. B. Johnson, *Biochemistry, 20*:6103 (1981).
126. H. K. Majumder, U. Maitra, and M. Rosenberg, *Proc. Natl. Acad. Sci. USA, 76*:5110 (1979).
127. B. Pace and N. R. Pace, *Anal. Biochem., 107*:128 (1980).
128. H. Schott, H. Schrade, and H. Watzlawick, *J. Chromatogr., 285*:343 (1984).
129. F. N. Verkleij, P. de Vries, and D. Peters, *J. Gen. Virol., 58*:329 (1982).
130. R. K. Tripathi and G. Kahl, *Biochem. Biophys. Res. Commun., 106*:1218 (1982).
131. R. K. Tripathi and G. Kahl, *Plant Cell Physiol., 23*:1101 (1982).
132. A. Konieczny and A. B. Legocki, *Acta Biochim. Pol., 28*:83 (1981).
133. C. R. Astell and M. Smith, *J. Biol. Chem., 246*:1944 (1971).
134. C. R. Astell and M. Smith, *Biochemistry, 11*:4114 (1972).
135. C. R. Astell, M. T. Doel, A. P. Jahnke, and M. Smith, *Biochemistry, 12*:5068 (1973).
136. S. Gillam, K. Waterman, and M. Smith, *Necleic Acids Res., 2*:625 (1975).
137. H. Schott, *J. Chromatogr., 187*:119 (1980).
138. H. Schott, *J. Chromatogr., 237*:429 (1982).
139. H. Schott and H. Watzlawick, *J. Chromatogr., 243*:57 (1982).

140. T. Ueda, K. Miura, S. Kobayashi, T. Fujiwara, and M. Yoshida, *Nucleic Acids Res. Symp. Ser. No. 6*:169 (1979).
141. J. Langer and Z. Kornetka, *Pol. J. Chem.*, *52*:1303 (1978).
142. D. Zuidema, F. M. van den Berg, and R. A. Flavell, *Nucleic Acids Res.*, *5*:2471 (1978).
143. V. Z. Tarantul, V. A. Lipasova, and K. G. Gazaryan, *Dokl. Akad. Nauk, SSSR*, *224*:719 (1976).
144. L. Yu. Frolova, Ya. Meldrajs, and L. L. Kisselev, *Dokl. Akad. Nauk. SSSR*, *225*:1442 (1975).
145. C. Coutelle, P. Ioannou, and R. Williamson, *Gene*, *3*:113 (1978).
146. A. P. Ryskov, N. E. Maleeva, and S. A. Limborska, *Gene*, *3*:81 (1978).
147. J. P. M. Sanders and P. Borst, *Mol. Gen. Genet.*, *157*:263 (1977).
148. A. P. Ryskov, G. N. Enikolopov, T. V. Vygodina, N. N. Dobbert, and G. P. Georgiev, *Dokl. Akad. Nauk. SSSR*, *232*:706 (1977).
149. B. M. Tyler and J. M. Adams, *Gene*, *10*:147 (1980).
150. K. Hayashi, *Gene*, *11*:109 (1980).
151. R. A. Owens and T. O. Diener, *Science*, *213*:670 (1981).
152. H. Sano, H. D. Rover, and R. Sager, *Proc. Natl. Acad. Sci. USA*, *77*:3581 (1980).
153. J. Iqbal, *Planta*, *134*:1 (1977).
154. J. G. J. Bauman, J. Wiegant, and P. van Duijn, *Histochemistry*, *73*:181 (1981).
155. M. Zeichner and D. Breitkreutz, *Arch. Biochem. Biophys.*, *188*:410 (1978).
156. J. Dennis and R. Kisilevsky, *Biochim. Biophys. Acta*, *561*:421 (1979).
157. E. Egyházi and A. Ossoinak, *Anal. Biochem.*, *92*:280 (1979).
158. S. G. Arsenyan, T. A. Avdonina, A. Laving, M. Saarma, and L. L. Kisseley, *Gene*, *11*:97 (1981).
159. A. J. Flavell, S. W. Ruby, J. J. Toole, R. E. Roberts, and G. M. Rubin, *Proc. Natl. Acad. Sci. USA*, *77*:7107 (1980).
160. K. Hemminki, *Hoppe-Seyler's Z. Physiol. Chem.*, *357*:1 (1976).
161. N. S. Grigoryan, *Mol. Biol.*, *14*:640 (1980).
162. A. Shapira and M. A. Billeter, *J. Biol. Chem.*, *253*:8400 (1978).
163. R. J. Colonno and H. O. Stone, *J. Virol.*, *17*:737 (1976).

164. N. L. Varich, I. S. Lukashevich, and N. V. Kaverin, *J. Viral. 18*:111 (1976).
165. J. N. Ihle, K.-L. Lee, and F. T. Kenney, *J. Biol. Chem., 249*:38 (1974).
166. R. G. Smith, L. Donehower, R. C. Gallo, and D. H. Gillespie, *J. Virol., 17*:287 (1976).
167. P. R. Etkind and R. M. Krug, *J. Virol., 16*:1464 (1975).
168. T. J. R. Harris and F. Brown, *J. Gen. Virol., 33*:493 (1976).
169. D. J. Rowlands, T. J. R. Harris, and F. Brown, *J. Virol., 26*:335 (1978).
170. A. Margotat, J. Laplane, M.-R. Pisano, and J. Nicoli, *Ann. Microbiol., 127*:243 (1976).
171. A. Leone, A. J. Shatkin, and R. Cancedda, *FEBS Lett., 100*:103 (1979).
172. F. Harada, N. Kato, and H. Hoshino, *Nucleic Acids Res., 7*:909 (1979).
173. R. Klemenz, M. Reinhardt, and H. Diggelmann, *Mol. Biol. Rep., 7*:123 (1981).
174. P. R. Srinivasan, M. Ramanarayanan, and E. Rabbani, *Proc. Natl. Acad. Sci. USA, 72*:2910 (1975).
175. P. Kerjan and J. Szulmajster, *Biochem. Biophys. Res. Commun., 93*:201 (1980).
176. Y. Gopalakrishna, D. Langley, and N. Sarkár, *Nucleic Acids Res., 9*:3545 (1981).
177. Y. Gopalakrishna and N. Sarkár, *Biochemistry, 21*:2724 (1982).
178. H. Ragg, J. Schroeder, and K. Hahlbrock, *Mol. Biol. Rep., 2*:119 (1975).
179. G. Trapy and R. Esnault, *Phytochemistry, 17*:1859 (1978).
180. R. Esnault, G. Trapy, and R. B. van Huystee, *Eur. J. Biochem., 60*:83 (1975).
181. J. L. Nichols and L. Welder, *Biochim. Biophys. Acta, 652*:99 (1981).
182. L. M. Dubroff, *Proc. Natl. Acad. Sci. USA, 74*:2217 (1977).
183. V. M. Kish and T. Pederson, *Proc. Natl. Acad. Sci. USA, 74*:1426 (1977).
184. R. H. Burdon, A. Shenkin, J. T. Douglas, and E. J. Smillie, *Biochim. Biophys. Acta, 474*:254 (1977).
185. D. Rosen and M. Edelman, *Eur. J. Biochem., 63*:525 (1976).
186. F. J. Hendler, G. Padmanaban, J. Patzer, R. Rayan, and M. Rabinowitz, *Nature, 258*:357 (1975).

187. G. Padmanaban, F. Hendler, J. Patzer, R. Rayan, and M. Rabinowitz, *Proc. Natl. Acad. Sci. USA, 72*:4293 (1975).
188. P. Cantatore, C. De Giorgi, and C. Saccone, *Biochem. Biophys. Res. Commun., 70*:431 (1976).
189. J. R. Hammett and F. R. Katterman, *Biochemistry, 14*:4375 (1975).
190. R. J. Glazer, *Biochim. Biophys. Acta, 418*:160 (1976).
191. P. E. Mirkes and B. McCalley, *J. Bacteriol., 125*:174 (1976).
192. A. J. Jaworski, *Arch. Biochem. Biophys., 173*:201 (1976).
193. J. J. Monahan, S. E. Harris, and B. W. O'Malley, *J. Biol. Chem., 251*:3738 (1976).
194. L. J. Rosenthal, *Nucleic Acids Res., 3*:661 (1976).
195. L. J. Rosenthal, C. Salomon, and R. Weil, *Nucleic Acids Res., 3*:1167 (1976).
196. J. DeLarco, A. Abramowitz, K. Bromwell, and G. Guroff, *J. Neurochem., 24*:215 (1975).
197. J. DeLarco, S. Nakagawa, A. Abramowitz, K. Bromwell, and G. Guroff, *J. Neurochem., 25*:131 (1975).
198. W. Berthold and L. Lim, *Biochem. J., 154*:517 (1976).
199. J. B. Mahony and I. R. Brown, *J. Neurochem., 25*:503 (1975).
200. J. W. Bynum and E. Volkin, *Anal. Biochem., 107*:406 (1980).
201. F. J. S. Lara and M. C. Okretic, *Biochem. J., 151*:575 (1975).
202. H. Grosjean, C. Takada, and J. Petre, *Biochem. Biophys. Res. Commun., 53*:882 (1973).
203. M. Edmonds, M. H. Vaughan, and H. Nakazato, *Proc. Natl. Acad. Sci. USA, 68*:1336 (1971).
204. G. R. Molloy and J. E. Darnell, *Biochemistry, 12*:2324 (1973).
205. G. Brawerman, J. Mendecki, and S. Y. Lee, *Biochemistry, 11*:637 (1972).
206. G. Brawerman, *Methods Enzymol., 30*:605 (1974).
207. J. M. Rosen, S. L. C. Woo, J. W. Holder, A. R. Means, and B. W. O'Malley, *Biochemistry, 14*:69 (1975).
208. J. Gorski, M. R. Morrison, C. G. Merkel, and J. B. Lingrel, *J. Mol. Biol., 86*:363 (1974).
209. P. A. Kitos, T. Fuller, G. S. King, and R. T. Hersh, *Biochim. Biophys. Acta, 353*:362 (1974).
210. P. A. Kitos and H. Amos, *Biochemistry, 12*:5086 (1973).

211. U. Lindberg and T. Persson, *Methods Enzymol.*, *34*:496 (1974).
212. J. DeLarco and G. Guroff, *Biochem. Biophys. Res. Commun.*, *50*:486 (1973).
213. J.-L. Ochoa, J. Kempf, and J. M. Egly, *Int. J. Biol. Macromol.*, *2*:33 (1980).
214. A. E. Berman, N. P. Gornaeva, and V. I. Mazurov, *Biokhimiya*, *43*:1830 (1978).
215. J. F. B. Mercer and H. Naora, *J. Chromatogr.*, *114*:115 (1975).
216. J. A. Bantle, I. H. Maxwell, and W. E. Hahn, *Anal. Biochem.*, *72*:413 (1976).
217. R. E. Pemberton, P. Liberti, and C. Baglioni, *Anal. Biochem.*, *66*:18 (1975).
218. L. T. Auger and G. F. Saunders, *Anal. Biochem.*, *79*:338 (1977).
219. S. Ohi and J. Short, *J. Appl. Biochem.*, *2*:398 (1980).
220. R. Grotha, *Biochem. Physiol. Pflanz.*, *170*:273 (1976).
221. J. M. Taylor, *Annu. Rev. Biochem.*, *48*:681 (1979).
222. J. Kates, *Cold Spring Harbor Symp. Quant. Biol.*, *35*:743 (1970).
223. U. Lindberg, T. Persson, and L. Philipson, *J. Virol.*, *10*:909 (1972).
224. H. Person, M. Perricaudet, A. Tolun, L. Philipson, and U. Pettersson, *J. Biol. Chem.*, *254*:7999 (1979).
225. J. P. Dudley and H. E. Varmus, *J. Virol.*, *39*:207 (1981).
226. S. G. Siddell, *Eur. J. Biochem.*, *92*:621 (1978).
227. L. Trombley and C.-S. Wang, *Microbiol. Immunol.*, *23*:629 (1979).
228. L. E. Holland, K. P. Anderson, J. R. Stringer, and E. K. Wagner, *J. Virol.*, *31*:447 (1979).
229. E. K. Wagner, R. J. Frink, and K. P. Anderson, *J. Virol.*, *39*:559 (1981).
230. C. S. McLaughlin, J. R. Warner, M. Edmonds, H. Nakazato, and M. H. Vaughan, *J. Biol. Chem.*, *248*:1466 (1973).
231. M. J. Holland, G. L. Hager, and W. J. Rutter, *Biochemistry*, *16*:8 (1977).
232. R. W. Haylock and E. A. Bevan, *Curr. Genet.*, *4*:181 (1981).
233. A. Ron, O. Horovitz, and I. Sarov, *J. Mol. Evol.*, *8*:137 (1976).
234. C. M Palatnik, R. V. Storti, and A. Jacobson, *J. Mol. Biol.*, *150*:389 (1981).

235. R. E. Gray and A. R. Cashmore, *J. Mol. Biol.*, *108*:595 (1976).
236. E. M. Tobin and A. D. Klein, *Plant Physiol.*, *56*:88 (1975).
237. D. A. Stuart, T. J. Mozer, and J. E. Varner, *Biochem. Biophys. Res. Commun.*, *105*:582 (1982).
238. J. Pasternak, R. J. Winkfein, and M. A. Fernando, *Mol. Biochem. Parasitol.*, *3*:133 (1981).
239. J. M. Sierra, W. Filipowicz, and S. Ochoa, *Biochem. Biophys. Res. Commun.*, *69*:181 (1976).
240. A. J. Ouellette, D. Frederick, and R. A. Matt, *Biochemistry*, *14*:4361 (1975).
241. A. J. Ouellette, A. Kumar, and R. A. Malt, *Biochim. Biophys. Acta*, *425*:384 (1976).
242. A. J. Minty and F. Gros, *J. Mol. Biol.*, *139*:61 (1980).
243. G. P. Siegal, C. P. Hodgson, P. K. Elder, L. S. Stoddard, and M. J. Getz, *J. Cell. Physiol.*, *103*:417 (1980).
244. R. B. Moffett and D. Doyle, *Biochim. Biophys. Acta*, *652*:177 (1981).
245. G. S. Cox, B. D. Weintraub, S. W. Rosen, and E. S. Maxwell, *J. Biol. Chem.*, *251*:1723 (1976).
246. M. Adesnik, M. Salditt, W. Thomas, and J. E. Darnell, *J. Mol. Biol.*, *71*:21 (1972).
247. H. Nakazato and M. Edmonds, *J. Biol. Chem.*, *247*:3365 (1972).
248. E. L. Korwek, H. Nakazato, S. Venkatesan, and M. Edmonds, *Biochemistry*, *15*:4643 (1976).
249. W. I. Murphy and G. Attardi, *Biochem. Biophys. Res. Commun.*, *74*:291 (1977).
250. G. R. Molloy, *J. Biol. Chem.*, *255*:10375 (1980).
251. M. Edmonds and W. M. Wood, *Biochemistry*, *20*:5359 (1981).
252. M. Edmonds and M. G. Caramela, *J. Biol. Chem.*, *244*:1314 (1969).
253. R. H. Stevens and A. R. Williamson, *Nature*, *239*:143 (1972).
254. K. A. Abraham, T. S. Eikhom, R. M. Dowben, and O. Garatun-Tjeldsto, *Eur. J. Biochem.*, *65*:79 (1976).
255. F. W. Hirsch, K. N. Nall, W. H. Spohn, and H. Busch, *Proc. Natl. Acad. Sci. USA*, *75*:1736 (1978).
256. P. K. Katinakis and R. H. Burdon, *Biochim. Biophys. Acta*, *653*:27 (1981).
257. R. E. Paque, *Cell Immunol.*, *30*:332 (1977).
258. B. J. Thiele, B. Papies, and C. Coutelle, *Acta Biol. Med. Ger.*, *37*:1317 (1978).

259. B. J. Thiele, Ch. Coutelle, H. D. Hunger, and A. P. Ryskov, *Acta Biol. Med. Ger.*, *37*:1331 (1978).
260. Ph. Tarroux, *Biochimie*, *57*:757 (1975).
261. L. Philipson, R. Wall, G. Glickman, and J. E. Darnell, *Proc. Natl. Acad. Sci. USA*, *68*:2806 (1971).
262. R. S. S. Fraser, *Eur. J. Biochem.*, *60*:477 (1975).
263. A. J. P. Brown and N. Hardman, *J. Gen. Microbiol.*, *122*:143 (1981).
264. L. Barakett and D. T. N. Pillay, *Gerontology*, *28*:1 (1982).
265. H. Aviv and P. Leder, *Proc. Natl. Acad. Sci. USA*, *69*:1408 (1972).
266. A. Krystosek, M. L. Cawthon, and D. Kabat, *J. Biol. Chem.*, *250*:6077 (1975).
267. D. Kabat, *J. Biol. Chem.*, *250*:6085 (1975).
268. J. Gielen, H. Aviv, and P. Leder, *Arch. Biochem. Biophys.*, *163*:146 (1974).
269. U. Nudel, H. Soreq, and U. Z. Littauer, *Eur. J. Biochem.*, *64*:115 (1976).
270. P. Nokin, G. Huez, G. Marbaix, A. Burny, and H. Chantrenne, *Eur. J. Biochem.*, *62*:509 (1976).
271. A. P. Ryskov, O. V. Tokarskaya, G. P. Georgiev, C. Coutelle, and B. Thiele, *Nucleic Acids Res.*, *3*:1487 (1976).
272. G. M. Maniatis, F. Ramirez, A. Cann, P. A. Marks, and A. Bank, *J. Clin. Invest.*, *58*:1419 (1976).
273. T. Cheng and H. H. Kazazian, Jr., *J. Biol. Chem.*, *252*:1758 (1977).
274. R. K. Strair, A. I. Skoultchi, and D. A. Shafritz, *Cell*, *12*:133 (1977).
275. Yu. N. Baranov, T. N. Zabojkina, V. I. Dubovaya, V. Z. Tarantul, and K. G. Gazaryan, *Dokl. Akad. Nauk. SSSR*, *234*:955 (1977).
276. R. J. Crawford and J. R. E. Wells, *Biochemistry*, *17*:1591 (1978).
277. T. G. Wood and J. B. Lingrel, *J. Biol. Chem.*, *252*:457 (1977).
278. D. H. Hamer and P. Leder, *Nature*, *281*:35 (1979).
279. S. A. Neifakh, V. S. Gajtskhoki, N. A. Klimov, L. V. Puchkova, M. M. Shavlovskij, and A. L. Schwartzman, *Mol. Biol. Rep.*, *3*:235 (1977).
280. M. Beato and A. Nieto, *Eur. J. Biochem.*, *64*:15 (1976).
281. I. Schechter, *Proc. Natl. Acad. Sci. USA*, *70*:2256 (1973).
282. I. Schechter, *Biochem. Biophys. Res. Commun.*, *67*:228 (1975).

283. T. Honjo, D. Swan, M. Nau, B. Norman, S. Packman, F. Polsky, and P. Leder, *Biochemistry, 15*:2775 (1976).
284. N. J. Cowan, D. S. Secher, and C. Milstein, *Eur. J. Biochem., 61*:355 (1976).
285. C. Auffray and F. Rougeon, *Eur. J. Biochem., 107*:303 (1980).
286. C. Auffray and F. Rougeon, *Gene, 12*:77 (1980).
287. C. Auffray, R. Nageotte, J.-L. Sikorav, O. Heidmann, and F. Rougeon, *Gene, 13*:365 (1981).
288. J. R. Parnes, B. Velan, A. Felsenfeld, L. Ramanathan, U. Ferrini, E. Appella, and J. G. Seidman, *Proc. Natl. Acad. Sci. USA, 78*:2253 (1981).
289. S. M. Heywood, D. S. Kennedy, and A. J. Bester, *FEBS Lett., 53*:69 (1975).
290. M. Patrinou-Georgoulas and H. A. John, *Cell, 12*:491 (1977).
291. H. Boedtker, A. M. Frischauf, and H. Lehrach, *Biochemistry, 15*:4765 (1976).
292. A. M. Frischauf, H. Lehrach, C. Rosner, and H. Boedtker, *Biochemistry, 17*:3243 (1978).
293. D. Breitkreutz, L. Diaz De Leon, L. Paglia, M. Zeichner, J. Wilczek, and R. Stern, *Biochim. Biophys. Acta, 517*:349 (1978).
294. R. Kaufmann, A. Belayew, B. Nusgens, C. M. Lapiere, and J. E. Gielen, *Eur. J. Biochem., 106*:593 (1980).
295. R. J. Rokowski, J. Sheehy, and K. R. Cutroneo, *Arch. Biochem. Biophys., 210*:74 (1981).
296. A. J. M. Berns, H. Bloemendal, S. J. Kaufman, and I. M. Verma, *Biochem. Biophys. Res. Commun., 52*:1013 (1973).
297. A. J. M. Vermorken, J. M. C. H. Hilderink, W. J. M. Van De Ven, and H. Bloemendal, *Biochim. Biophys. Acta, 454*:447 (1976).
298. L. H. Cohen, L. W. Westerhuis, D. P. Smits, and H. Bloemendal, *Eur. J. Biochem., 89*:251 (1978).
299. M. B. Mathews, M. Osborn, A. J. M. Berns, and H. Bloemendal, *Nature, 236*:5 (1972).
300. P. Cashion, H. Notman, G. Sathe, T. Cadger, K. Porter, and E. Jay, *Nucleic Acids Res., 4*:2593 (1977).
301. P. Cashion, G. Sathe, A. Javed, and J. Kuster, *Nucleic Acids Res., 8*:1167 (1980).
302. W. Burnett, K. Yoon, and J. Rosenbloom, *Biochem. Biophys. Res.. Commun., 99*:364 (1981).
303. J. Schweizer and K. Goerttler, *Eur. J. Biochem., 112*:243 (1980).

304. R. E. Rhoads, *J. Biol. Chem.*, *250*:8088 (1975).
305. S. L. C. Woo, R. G. Smith, A. R. Means, and B. W. O'Malley, *J. Biol. Chem.*, *251*:3868 (1976).
306. S. L. C. Woo, T. Chandra, A. R. Means, and B. W. O'Malley, *Biochemistry*, *16*:5670 (1977).
307. R. E. Rhoads and G. M. Hellmann, *J. Biol. Chem.*, *253*:1687 (1978).
308. J. M. Rosen, *Biochemistry*, *15*:5263 (1976).
309. L.-M. Houdebine and P. Gaye, *Eur. J. Biochem.*, *63*:9 (1976).
310. J. M. Taylor and T. P. H. Tse, *J. Biol. Chem.*, *251*:7461 (1976).
311. P. K. Chakrabartty and P. K. Qasba, *Nucleic Acids Res.*, *4*:2065 (1977).
312. G. J. Dimitriadis, *Eur. J. Biochem.*, *118*:255 (1981).
313. B. A. Larkins, R. A. Jones, and C. Y. Tsai, *Biochemistry*, *15*:5506 (1976).
314. I. M. Evans, R. R. D. Croy, P. Brown, and D. Boulter, *Biochim. Biophys. Acta*, *610*:81 (1980).
315. D. J. Shapiro, H. J. Baker, and D. T. Stitt, *J. Biol. Chem.*, *251*:3105 (1976).
316. R. G. Deeley, J. I. Gordon, A. T. H. Burns, K. P. Mullinix, M. Binastein, and R. F. Goldberger, *J. Biol. Chem.*, *252*:8310 (1977).
317. J. Mahony, I. Brown, G. Labourdette, and A. Marks, *Eur. J. Biochem.*, *67*:203 (1976).
318. J. B. Fagan, K. M. Yamada, B. de Crombrugghe, and I. Pastan, *Nucleic Acids Res.*, *6*:3471 (1979).
319. A. Fuse, H. Heremans, G. Opdenakker, and A. Billiau, *Biochem. Biophys. Res. Commun.*, *105*:1309 (1982).
320. H. Weening, A. Fuse, G. Opdenakker, J. Van Damme, M. De Ley, and A. Billiau, *Biochem. Biophys. Res. Commun.*, *104*:6 (1982).
321. S. M. Tilghman, L. M. Fisher, L. Reshef, F. J. Ballard, and R. W. Hanson, *Biochem. J.*, *156*:619 (1976).
322. P. B. Iynedjian and R. W. Hanson, *Eur. J. Biochem.*, *90*:123 (1978).
323. D. Sagher, H. Grosfeld, and M. Edelman, *Proc. Natl. Acad. Sci. USA*, *73*:722 (1976).
324. V. V. Adler, V. L. Mechitov, and V. S. Shapot, *Dokl. Akad. Nauk. SSSR*, *233*:719 (1977).
325. P. Madhusudanan Nair and K. K. Ussuf, *Indian J. Exp. Biol.*, *17*:460 (1979).

326. J. M. Boss, S. Gillam, R. S. Zitomer, and M. Smith, *J. Biol. Chem.*, *256*:12958 (1981).
327. K. Sogawa, Y. Ichihara, K. Takahashi, Y. Fujii-Kuriyama, and M. Muramatsu, *J. Biol. Chem.*, *256*:12561 (1981).
328. K. Tsutsumi and K. Ishikawa, *Biochem. Biophys. Res. Commun.*, *100*:407 (1981).
329. G. Vassart, H. Brocas, P. Nokin, and J. E. Dumont, *Biochim. Biophys. Acta*, *324*:575 (1973).
330. G. Vassart, H. Brocas, R. Lecocq, and J. E. Dumont, *Eur. J. Biochem.*, *55*:15 (1975).
331. G. Vassart, S. Refetoff, H. Brocas, C. Dinsart, and J. E. Dumont, *Proc. Natl. Acad. Sci. USA*, *72*:3839 (1975).
332. J. R. Dugiud, D. F. Steiner, and W. L. Chick, *Proc. Natl. Acad. Sci. USA*, *73*:3539 (1976).
333. M. Mevarech, B. E. Noyes, and K. L. Agarwal, *J. Biol. Chem.*, *254*:7472 (1979).
334. M. A. Shupnik and J. Gorski, *Mol. Cell. Endocrinol.*, *17*:181 (1980).
335. J. H. Nilson, A. R. Thomason, S. Horowitz, N. L. Sasavage, J. Blenis, R. Albers, W. Salser, and F. M. Rottman, *Nucleic Acids Res.*, *8*:1561 (1980).
336. W. L. Miller, J.-P. Thirion, and J. A. Martial, *Endocrinology*, *107*:851 (1980).
337. W. L. Miller, J. A. Martial, and J. D. Baxter, *J. Biol. Chem.*, *225*:7521 (1980).
338. L. Gedamu and G. H. Dixon, *J. Biol. Chem.*, *251*:1455 (1976).
339. K. Iatrou and G. H. Dixon, *Cell*, *10*:433 (1977).
340. R. G. Levenson and K. B. Marcu, *Cell*, *9*:311 (1976).
341. G. Childs, S. Levy, and L. H. Kedes, *Biochemistry*, *18*:208 (1979).
342. P. A. Krieg, A. J. Robins, M. J. Gait, R. C. Titmas, and J. R. E. Wells, *Nucleic Acids Res.*, *10*:1495 (1982).
343. A. Mutsukage, E. W. Bohn, and S. H. Wilson, *Biochemistry*, *14*:1006 (1975).
344. R. Mukai and Y. Iida, *Biochem. Biophys. Res. Commun.*, *54*:134 (1973).
345. P. Plevani, A. M. Albertini, A. Galizzi, A. Adamoli, G. Mastromei, S. Riva, and G. Cassani, *Nucleic Acids Res.*, *4*:603 (1977).
346. T. Kristensen and J. Holtlund, *Eur. J. Biochem.*, *70*:441 (1976).
347. B. L. Gerwin and J. B. Milstein, *Proc. Natl. Acad. Sci. USA*, *69*:2599 (1972).

348. D. M. L. Livingston, E. M. Scolnick, W. P. Parks, and G. J. Todaro, *Proc. Natl. Acad. Sci. USA*, *69*:393 (1972).
349. S. L. Marcus, M. J. Modak, and L. F. Cavalieri, *J. Virol.*, *14*:853 (1974).
350. D. P. Grandgenett and H. M. Rho, *J. Virol.*, *15*:526 (1975).
351. P. K. Majumdar and V. A. Vipparti, *FEBS Lett.*, *109*:31 (1980).
352. H. Lehrach and K. H. Scheit, *Hoppe-Seyler's Z. Physiol. Chem.*, *353*:731 (1972).
353. S. C. Weatherford, L. S. Weisberg, D. T. Achord, and D. Apirion, *Biochem. Biophys. Res. Commun.*, *49*:1307 (1972).
354. M. N. Thang, J. L. Drocourt, M. K. Chelbi-Alix, D. C. Thang, J. Lubochinsky, A. Ruet, A. Sentenac, J. Gangloff, and G. Dirheimer, in *Affinity Chromatography and Molecular Interactions* (J. M. Egly, Ed.), INSERM Symposia Series, Vol. 86, Paris, 1979, p. 303.
355. S. Okamura, F. Crane, H. A. Messner, and T. W. Mak, *J. Biol. Chem.*, *253*:3765 (1978).
356. P. Zabel, L. Dorsses, K. Wernars, and A. van Kammen, *Nucleic Acids Res.*, *9*:2433 (1981).
357. G. Brun, P. Vannier, I. Scovassi, and J.-C. Callen, *Eur. J. Biochem.*, *118*:407 (1981).
358. O. D. Nelidova and L. L. Kiselev, *Mol. Biol.*, *2*:60 (1968).
359. H. Hayashi, *J. Biochem.*, *74*:203 (1973).
360. S. Bartkowiak and J. Pawelkiewicz, *Biochim. Biophys. Acta*, *272*:137 (1972).
361. W. Kedzierski and J. Pawelkiewicz, *Bull. Acad. Pol. Sci., Ser. Sci. Biol.*, *25*:637 (1977).
362. G. Sivakumar Swamy and D. T. N. Pillay, *Pflanzenphysiologie*, *93*:403 (1979).
363. V. Natarajan and K. P. Gopinathan, *Indian J. Biochem. Biophys.*, *17*:330 (1980).
364. G. Fayat, M. Fromant, D. Kahn, and S. Blanquet, *Eur. J. Biochem.*, *78*:333 (1977).
365. M. Fromant, G. Fayat, P. Laufer, and S. Blanquet, *Biochimie*, *63*:541 (1981).
366. J. C. Schabort, *J. Chromatogr.*, *73*:253 (1972).
367. L. Jervis and N. M. Pettit, *J. Chromatogr.*, *97*:33 (1974).
368. J. Veleminský, J. Svachulová, and J. Satava, *Nucleic Acids Res.*, *8*:1373 (1980).
369. J. Rouvière-Yaniv and F. Gros, *Proc. Natl. Acad. Sci. USA*, *72*:3428 (1975).

370. L. Reichard, Ph.D. dissertation, Stanford University, 1972.
371. R. Haselkorn and J. Rouvière-Yaniv, *Proc. Natl. Acad. Sci. USA*, *73*:1917 (1976).
372. N. N. G. Tsao and R. E. Pearlman, *Can. J. Biochem.*, *60*: 398 (1982).
373. B. Kallin, J. Roubal, J. Luka, and G. Klein, *Virology*, *113*:285 (1981).
374. J. C. Bearden, Jr., *J. Biochem. Biophys. Methods*, *2*:37 (1980).
375. H. Weideli and W. J. Gehring, *Eur. J. Biochem.*, *104*:5 (1980).
376. H. Weideli, Ch. Brack, and W. J. Gehring, *Proc. Natl. Acad. Sci. USA*, *77*:3773 (1980).
377. H. Weideli, P. Schedl, S. Artavanis-Tsakonas, R. Steward, R. Yuan, and W. J. Gehring, *Cold Spring Harbor Symp. Quant. Biol.*, *42*:693 (1978).
378. P. Nehls and M. Renz, *Anal. Biochem.*, *107*:124 (1980).
379. J. M. Bogden, A. Eastman, and E. Bresnick, *Nucleic Acids Res.*, *9*:3089 (1981).
380. G. G. Carmichael, K. Weber, A. Niveleau, and A. J. Wahba, *J. Biol. Chem.*, *250*:3607 (1975).
381. S. K. Jain, M. G. Pluskal, and S. Sarkar, *FEBS Lett.*, *97*: 84 (1979).
382. B. Axelsson, R. Ohlsson, A. Deutsch, and B. Jergil, *Mol. Cell. Biochem.*, *15*:67 (1977).
383. B. Setyono, M. Grossmann, and J.-P. Liautard, *Biochimie*, *59*:43 (1977).
384. J. P. Liautard, D. Tromm, and K. Kohler, *Z. Naturforsch. Ser. C*, *32*:419 (1977).
385. D. Irwin, A. Kumar, and R. A. Malt, *Cell*, *4*:157 (1975).
386. U. Lindberg and B. Sundquist, *J. Mol. Biol.*, *86*:451 (1974).
387. A. T. H. Burns and R. Williamson, *Nucleic Acids Res.*, *2*: 2251 (1975).
388. W. R. Jeffery and C. Peters, *Mol. Biol. Rep.*, *3*:379 (1977).
389. H. Slegers, E. de Herdt, and M. Kondo, *Eur. J. Biochem.*, *117*:111 (1981).
390. P. E. Mirkes, *J. Bacteriol.*, *131*:240 (1977).
391. G. D. Sinclair and G. H. Dixon, *Biochemistry*, *21*:1869 (1982).
392. C. H. Darnbrough and P. J. Ford, *Eur. J. Biochem.*, *113*: 415 (1981).

393. P. de Meyer, E. de Herdt, M. Kondm, and H. Slegers, *J. Biochem. Biophys. Methods, 2*:311 (1980).
394. A. M. Gressner and E. van de Leur, *Biochim. Biophys. Acta, 608*:459 (1980).
395. M. Yukioka and K. Omori, *FEBS Lett., 75*:217 (1977).
396. N. Ulbrich and I. G. Wool, *J. Biol. Chem., 253*:9049 (1978).
397. N. Ulbrich, A. Lin, K. Todokoro, and I. G. Wool, *J. Biol. Chem., 255*:797 (1980).
398. N. Ulbrich, I. G. Wool, E. Ackerman, and P. B. Sigler, *J. Biol. Chem., 255*:7010 (1980).
399. R. M. Elliott, A. N. Davison, and L. Lim, *Biochem. J., 190*:215 (1980).
400. S. K. Jain, R. K. Roy, M. G. Pluskal, D. E. Croall, C. Guha, and S. Sarkar, *Mol. Biol. Rep., 5*:79 (1979).
401. S. K. Jain and S. Sarkar, *Biochemistry, 18*:745 (1979).
402. A. Alonso, H. Winter, and L. Krieg, *Biochim. Biophys. Acta, 652*:72 (1981).
403. R. K. Roy, A. S. Lau, H. N. Munro, B. Surendra Baliga, and S. Sarkar, *Proc. Natl. Acad. Sci. USA, 76*:1751 (1979).
404. D. B. Sittman, R. A. Graves, and W. F. Marzluff, *Proc. Natl. Acad. Sci. USA, 80*:1849 (1983).
405. J. A. Hellerman and D. A. Shafritz, *Proc. Natl. Acad. Sci. USA, 72*:1021 (1975).
406. R. Kaempfer, R. Hollander, W. Arams, and R. Israeli, *Proc. Natl. Acad. Sci. USA, 75*:209 (1978).
407. O. Nygård, P. Westermann, and T. Hultin, *Biochim. Biophys. Acta, 608*:196 (1980).
408. S. M. Heywood and D. S. Kennedy, *Arch. Biochem. Biophys., 192*:270 (1979).
409. T. M. Jovin and A. Kornberg, *J. Biol. Chem., 243*:250 (1968).
410. N. R. Cozzarelli, N. E. Melechen, T. M. Jovin, and A. Kornberg, *Biochem. Biophys. Res. Commun., 28*:578 (1967).
411. A. Panet and H. G. Khorana, *J. Biol. Chem., 249*:5213 (1974).
412. P. Venetianer and P. Leder, *Proc. Natl. Acad. Sci. USA, 71*:3892 (1974).
413. S. Levy and H. Aviv, *Biochemistry, 15*:1844 (1976).
414. C. R. Thrash and R. T. Schimke, *J. Biol. Chem., 252*:5615 (1977).
415. F. W. Hirsch, K. N. Nall, L. C. Hayes, K. S. Raju, W. H. Spohn, and H. Busch, *Cancer Res., 37*:3694 (1977).

416. M. P. Vitek, S. G. Kreissman, and R. H. Gross, *Nucleic Acids Res.*, *9*:1191 (1981).
417. I. E. Scheffler and C. C. Richardson, *J. Biol. Chem.*, *247*:5736 (1972).
418. R. Roychoudhury, S. Kühn, H. Schott, and H. Kössel, *FEBS Lett.*, *50*:140 (1975).
419. C. Saxinger, C. Ponnamperuma, and C. Woese, *Nature, New Biol.*, *234*:172 (1971).
420. H. Eckstein, H. Schott, and E. Bayer, *Hoppe-Seyler's Z. Physiol. Chem.*, *356*:91 (1975).
421. H. Schott, H. Eckstein, I. Gatfield, and E. Bayer, *Biochemistry*, *14*:5541 (1975).
422. H. Eckstein, H. Schott, and E. Bayer, *Biochim. Biophys. Acta*, *432*:1 (1976).
423. J. T. Shapiro, B. S. Stannard, and G. Felsenfeld, *Biochemistry*, *8*:3233 (1969).
424. M. Leng and G. Felsenfeld, *Proc. Natl. Acad. Sci. USA*, *56*:1325 (1966).
425. K. Wehling, H.-A. Arfmann, K.-H. C. Standke, and K. G. Wagner, *Nucleic Acids Res.*, *2*:799 (1975).
426. H. Schott and H. Eckstein, *Hoppe-Seyler's Z. Physiol. Chem.*, *360*:1689 (1979).
427. H. Schott and H. Eckstein, *Eur. J. Biochem.*, *104*:79 (1980).
428. H. Schott and H. Eckstein, in *Structure and Activity of Natural Peptides* (W. Voelter and G. Weitzel, Eds.), Walter de Gruyter, Berlin, 1981, p. 326.
429. S. R. Ayad and J. Blamire, *Biochem. Biophys. Res. Commun.*, *30*:207 (1968).
430. C. W. Helleiner, *Can. J. Biochem.*, *47*:1199 (1969). D. Jawis, R. Loeser, R. Herrlich, and R. Röschenthaler, *J. Chromatogr.*, *52*:158 (1970).
431. C. W. Helleiner, *J. Chromatogr.*, *72*:400 (1972).
432. S. R. Ayad and M. C. Parker, *J. Chromatogr.*, *82*:343 (1973).
433. H. Panusz, J. Bartkowiak, A. Krzywiec, M. Joustra, and A. Plucienniczak, *Nucleic Acids Res.*, *1*:1143 (1974).
434. P. P. Lau, S.-H. Yu, T. G. Spring, and H. B. Gray, Jr., *Biochim. Biophys. Acta*, *563*:313 (1979).
435. K. Horiuchi, D. Fujimoto, Y. Seto, and S. Sakakibara, *J. Biochem.*, *83*:261 (1978).
436. R.-K. Selander and A. de La Chapelle, *Acta Chem. Scand. B*, *30*:925 (1976).

437. K. B. Palter and B. M. Alberts, *J. Biol. Chem.*, *254*:11160 (1979).
438. J. Fuhrmann, B. A. thesis, Princeton University, 1974.
439. G. Herrick, *Nucleic Acids Res.*, *8*:3721 (1980).
440. K. H. Thanki, T. A. Beach, and H. W. Dickerman, *J. Biol. Chem.*, *253*:7744 (1978).
441. S. J. Wieland and T. O. Fox, *J. Steroid Biochem.*, *14*:409 (1981).
442. A. S. Gambaryan, I. A. Morozov, T. V. Venkstern, and A. A. Bayev, *Nucleic Acids Res.*, *6*:1001 (1979).
443. K. Yoshihara, T. Hashida, Y. Tanaka, N. Matsunami, A. Yamaguchi, and T. Kamiya, *J. Biol. Chem.*, *256*:3471 (1981).
444. A. E. Pegg and A. Jackson, *Chem. Biol. Interact.*, *12*:279 (1976).
445. W.-F. Fong and M. S. Fuchs, *Insect Biochem.*, *6*:123 (1976).
446. L. J. Holland and L. J. Wangh, *Nucleic Acids Res.*, *11*:3283 (1983).
447. H. Bünemann, P. Westhoff, and R. G. Herrmann, *Nucleic Acids Res.*, *10*:7163 (1982).
448. H. Bünemann, *Nucleic Acids Res.*, *10*:7181 (1983).
449. Yu. P. Zerov, A. V. Averyanov, K. A. Kollov, and L. S. Chesnokova, *Biokhimiya*, *48*:292 (1983).
450. C. Arus, M. V. Nogués, and C. M. Cuchillo, *J. Chromatogr.*, *237*:500 (1982)
451. L. Graeve, W. Goemann, P. Földi, and J. Kuppa, *Biochem. Biophys. Res. Commun.*, *107*:1559 (1982).
452. S. Nadin-Davis and V. A. Mezl, *Prep. Biochem.*, *12*:49 (1982).
453. J. N. Anderson and R. T. Schimke, *Cell*, *7*:331 (1976).
454. S. L. C. Woo, J. J. Monahan, and B. W. O'Malley, *J. Biol. Chem.*, *252*:5789 (1977).
455. R. A. Firtel and H. F. Lodish, *J. Mol. Biol.*, *79*:295 (1973).
456. C. M. Palatnik, R. V. Storti, and A. Jacobson, *J. Mol. Biol.*, *128*:371 (1979).
457. J. C. Alwine, D. J. Kemp, and G. R. Stark, *Proc. Natl. Acad. Sci. USA*, *74*:5350 (1977).
458. J. Ross, *J. Mol. Biol.*, *119*:21 (1978).
459. D. J. Shapiro and R. T. Schimke, *J. Biol. Chem.*, *250*:1759 (1975).
460. J. M. Taylor and R. W. Hamilton, in *Affinity Chromatography and Molecular Interactions* (J. M. Egli, Ed.), INSERM Symposia Series, Vol. 86, Paris, 1979, p. 265.

461. A. Leriche, D. Christophe, H. Brocas, and G. Vassart, *Anal. Biochem.*, *129*:249 (1983).
462. L. Hwang-Lee, J. Blamire, and S. F. Cottrell, *Anal. Biochem.*, *128*:47 (1983).
463. H.-E. Wilk, N. Kecskemethy, and K. P. Schäfer, *Nucleic Acids Res.*, *10*:7621 (1982).
464. J. Reiser, J. Renart, and G. R. Stark, *Biochem. Biophys. Res. Commun.*, *85*:1104 (1978).
465. Y. Kitagawa and E. Okuhara, *Anal. Biochem.*, *115*:102 (1981).
466. F. Harada and Y. Ikawa, *Nucleic Acids Res.*, *7*:895 (1979).
467. E. Mothes, *Arch. Geschwulstforsch.*, *52*:191 (1982).
468. T. Otal and V. Hari, *Virology*, *125*:118 (1983).
469. I. Hussain, N. Tsukagoshi, and S. Udaka, *J. Bacteriol.*, *151*:1162 (1982).
470. K. K. Ussuf and R. M. Nair, *Indian J. Biochem. Biophys.*, *18*:276 (1981).
471. J. R. Greenberg and B. Setyono, *Biol. Cell*, *41*:67 (1981).
472. J. L. Nichols and L. Welder, *Plant Sci. Lett.*, *29*:61 (1983).
473. A. Bernd, R. K. Zahn, A. Maidhof, and W. E. G. Müller, *Hoppe-Seyler's Z. Physiol. Chem.*, *363*:221 (1982).
474. G. R. Molloy and S. Johnson, *Prep. Biochem.*, *12*:77 (1982).
475. Z. Lassota, J. Michalik, M. Szyszko, and A. Krówczyńska, *Acta Biochim. Pol.*, *26*:83 (1979).
476. P. Z. Khasigov, V. F. Glazkov, A. A. Del'vig, D. A. Kuznetsov, and A. Ya. Nikolaev, *Biokhimiya*, *48*:179 (1983).
477. A. Dautigny, P. M. Alliel, J.-L. Nussbaum, and P. Jolles, *Biochem. Biophys. Res. Commun.*, *110*:432 (1983).
478. J. C. Wallace and M. Edmonds, *Proc. Natl. Acad. Sci. USA*, *80*:950 (1983).
479. I. Barahona, L. Galego, and C. Rodrigues-Pousada, *Eur. J. Biochem.*, *131*:171 (1983).
480. J. Rey-Campos, M. Rey-Mendez, and M. Freire, *Int. J. Biochem.*, *15*:155 (1983).
481. M. K. Kistler, R. E. Taylor, Jr., J. C. Kandala, and W. S. Kistler, *Biochem. Biophys. Res. Commun.*, *99*:1161 (1981).
482. J.-L. Bergé-Lefranc, G. Cartouzou, Y. Malthiéry, F. Perrin, B. Jarry, and S. Lissitzky, *Eur. J. Biochem.*, *120*:1 (1981).

483. H. V. Molgaard, M. Perucho, and A. Ruiz-Carrillo, *Nature, 283*:502 (1980).
484. A. J. Ouellette and C. P. Ordahl, *J. Biol. Chem., 256*: 5104 (1981).
485. K. Smith, P. Rosteck, Jr., and J. B. Lingrel, *Nucleic Acids Res.*, 5:105 (1978).
486. P. C. Brown and J. Papaconstantinou, *J. Biol. Chem., 254*:5177 (1979).
487. J. Guozhong, C. Yan, C. Jianwen, and L. Yaozhen, *Acta Gen. Sin., 8*:241 (1981).
488. M. Sellami, B. Rether, J. Gangloff, J.-P. Ebel, and J. Bonnet, *Nucleic Acids Res., 11*:3269 (1983).
489. Y. Okai, M. Yukioka, and A. Inoue, *FEBS Lett., 141*:102 (1982).
490. A. Srivastava and M. J. Modak, *Biochemistry, 19*:3270 (1980).
491. M. Sugiura, M. Suzuki, E. Ohtsuka, S. Nishikawa, H. Uemura, and M. Ikehara, *FEBS Lett., 97*:73 (1979).
492. N. A. Zykkova, G. A. Nevinsky, and O. I. Lavrik, *Mol. Biol. (Moscow), 16*:1165 (1982).
493. T. Ohta, M. Ogawa, M. Kurihara, T. Kitahara, and G. Kosaki, *Clin. Chim. Acta, 124*:51 (1982).
494. H. Horitsu, N. Hara, M. Tohgane, and K. Kawai, *Agric. Biol. Chem., 46*:1145 (1982).
495. R. César and W. G. Verly, *Eur. J. Biochem., 129*:509 (1983).
496. C. Bonne, P. Sautiere, M. Duguet, and A.-M. de Recondo, *J. Biol. Chem., 257*:2722 (1982).
497. T. Sarapuu and R. Villems, *Eur. J. Biochem., 124*:275 (1982).
498. D. Przybyl, M. M. Sikorski, and A. B. Legocki, *Acta Biochim. Pol., 28*:77 (1981).
499. A. S. Stepanov, K. V. Kandror, and S. M. Elizarov, *FEBS Lett., 141*:157 (1982).
500. F. Tamanoi and B. W. Stillman, *Proc. Natl. Acad. Sci. USA, 79*:2221 (1982).
501. W. Fischer, K.-H. Derwenskus, and M. Sprinzl, *Eur. J. Biochem., 125*:143 (1982).
502. I. A. Morozov, A. S. Gambaryan, T. N. Lvova, A. A. Nedospasov, and T. V. Venkstern, *Eur. J. Biochem., 129*: 429 (1982).
503. I. Estepa and A. Pestana, *Eur. J. Biochem., 132*:249 (1983).

504. P. P. Baskevitch, F. Vignon, C. Bousquet, and H. Rochefort, *Cancer Res.*, *43*:2290 (1983).
505. E. R. Mulvihill, J.-P. LePennec, and P. Chambon, *Cell*, *24*:621 (1982).
506. P. Arányi, *Eur. J. Biochem.*, *129*:549 (1983).
507. B. H. Robertson, D. O. Morgan, D. M. Moore, M. J. Grubman, J. Card, T. Fischer, G. Weddell, D. Dowbenko, and D. Yansura, *Virology*, *126*:614 (1983).
508. R. A. Cox and N. J. Smulian, *FEBS Lett.*, *155*:73 (1983).
509. J. J. Leary, D. J. Brigati, and D. C. Ward, *Proc. Natl. Acad. Sci. USA*, *80*:4045 (1983).
510. M. Altmann, N. Kaeufer, and H. von Doehren, *FEMS Microbiol. Lett.*, *18*:245 (1983).
511. H. Fujimoto and R. P. Erickson, *Biochem. Biophys. Res. Commun.*, *108*:1369 (1982).
512. M. Witt, R. Slomski, and M. Kurpisz, *Acta Biochim. Pol.*, *29*:311 (1982).
513. E. Caldiroli, G. Zocchi, and S. Cocucci, *Eur. J. Biochem.*, *131*:255 (1983).
514. T. Formosa, R. L. Burke, and B. M. Alberts, *Proc. Natl. Acad. Sci. USA*, *80*:2442 (1983).
515. G. Akusjärvi and U. Petterson, *Proc. Natl. Acad. Sci. USA*, *75*:5822 (1978).
516. D. J. McGeoch and N. T. Turnbull, *Nucleic Acids Res.*, *5*:4007 (1978).
517. V. C. Hieber, *Biochem. Biophys. Res. Commun.*, *104*:1271 (1982).
518. Y. Lin, *J. Biol. Chem.*, *254*:1422 (1979).
519. Y. Lin and D. J. Long, *Biochemistry*, *19*:1111 (1980).
520. T. J. R. Harris, P. A. Lowe, A. Lyons, P. G. Thomas, M. A. W. Eaton, T. A. Millican, T. P. Patel, C. C. Bose, N. H. Carey, and M. T. Doel, *Nucleic Acids Res.*, *10*:2177 (1982).
521. K. Sakimura, K. Araki, E. Kushiya, and Y. Takahashi, *J. Neurochem.*, *39*:366 (1982).
522. J. P. Kraus and L. E. Rosenberg, *Proc. Natl. Acad Sci. USA*, *79*:4015 (1982).
523. D. O. Irving and M. J. Howell, *Mol. Biochem. Parasitol.*, *4*:337 (1981).
524. R. F. Troxler, H. S. Belford, and F. G. Oppenheim, *Biochem. Biophys. Res. Commun.*, *111*:239 (1983).
525. D. Bartels and R. D. Thompson, *Nucleic Acids Res.*, *11*:2961 (1983).

526. J. A. Gurr, J. F. Catterall, and I. A. Kourides, *Proc. Natl. Acad. Sci. USA, 80*:2122 (1983).
527. S. Hinuma, H. Onda, K.-I. Naruo, Y. Ichimori, M. Koyama, and K. Tsukamoto, *Biochem. Biophys. Res. Commun., 109*:363 (1982).
528. K. Nishimori, Y. Kawaguchi, M. Hidaka, T. Uozumi, and T. Beppu, *J. Biochem., 90*:901 (1981).
529. M. Houghton, A. G. Stewart, S. M. Doel, J. S. Emtage, M. A. W. Eaton, J. C. Smith, T. P. Patel, H. M. Lewis, A. G. Porter, J. R. Birch, T. Cartwright, and N. H. Carey, *Nucleic Acids Res., 8*:1913 (1980).
530. E. Boel, J. Vuust, F. Norris, K. Norris, A. Wind, J. F. Rehfeld, and K. A. Marcker, *Proc. Natl. Acad. Sci. USA, 80*:2866 (1983).
531. A. K. Sood, D. Pereira, and S. M. Weissman, *Proc. Natl. Acad. Sci. USA, 78*:616 (1981).
532. J. C. Boothroyd and G. A. M. Cross, *Gene, 20*:281 (1982).
533. I. R. Lemischka, S. Farmer, V. R. Racaniello, and P. A. Sharp, *J. Mol. Biol., 151*:101 (1981).
534. K. B. Marcu, O. Valbuena, and R. P. Perry, *Biochemistry, 17*:1723 (1978).
535. S. Nakanishi, A. Inoue, T. Kita, S. Numa, A. C. Y. Chang, S. N. Cohen, J. Nunberg, and R. T. Schimke, *Proc. Natl. Acad. Sci. USA, 75*:6021 (1978).
536. S. Nakanishi, A. Inoue, T. Kita, M. Nakamura, A. C. Y. Chang, S. N. Cohen, and S. Numa, *Nature, 278*:423 (1979).
537. R. Palacios, D. Sullivan, N. Morris Summer, M. L. Kiely, and R. T. Schimke, *J. Biol. Chem., 248*:540 (1973).
538. P. J. Curtis and C. Weissman, *J. Mol. Biol.,* 1061 (1976).
539. A. K. Deshpande, B. Chatterjee, and A. K. Roy, *J. Biol. Chem., 254*:8937 (1979).
540. P. W. Tucker, K. B. Marcu, J. L. Slightom, and F. R. Blattner, *Science, 206*:1299 (1979).
541. J. H. J. Hoeijmakers, P. Borst, J. van den Burg, C. Weissmann, and G. A. M. Cross, *Gene, 8*:391 (1980).
542. N. Rave, R. Crkvenjakov, and H. Boedtker, *Nucleic Acids Res., 6*:3559 (1979).
543. P. K. Davies, G. H. Dixon, A. Simoncsits, and G. G. Brownlee, *Nucleic Acids Res., 7*:2323 (1979).
544. T. Yagura, T. Kozu, and T. Seno, *Biochem. Biophys. Res. Commun., 92*:1289 (1980).
545. S. R. Ayad and J. B. Delinassios, *J. Chromatogr., 130*: 396 (1977).

546. M. Marty, M. Mauchauffe, J. Poli, and C. J. Larsen, *Biochimie, 56*:659 (1974).
547. A. M. Kotin and N. P. Teryukova, *Vopr. Med. Khim., 22*: 712 (1976).
548. Yu. P. Zerov, *Biokhimiya, 41*:35 (1976).
549. N. C. Vamvakopoulos and I. A. Kourides, *Proc. Natl. Acad. Sci. USA, 76*:3809 (1979).
550. M. C. Vamvakopoulos, J. J. Monahan, and I. A. Kourides, *Proc. Natl. Acad. Sci. USA, 77*:3149 (1980).
551. B. Howard, S. L. Adams, M. E. Sobel, I. Pastan, and B. de Crombrugghe, *J. Biol. Chem., 253*:5869 (1978).
552. M. E. Sobel, T. Yamamoto, S. L. Adams, R. Di Lauro, V. E. Avvedimento, B. de Crombrugghe, and I. Pastan, *Proc. Natl. Acad. Sci. USA, 75*:5846 (1978).
553. S. L. C. Woo, J. M. Rosen, C. D. Liarakos, Y. C. Choi, H. Busch, A. R. Means, and B. W. O'Malley, *J. Biol. Chem., 250*:7027 (1975).
554. L. A. Mc Reynolds, J. F. Catterall, and B. W. O'Malley, *Gene, 2*:217 (1977).
555. L. A. Mc Reynolds, J. J. Monahan, D. W. Bendure, S. L. C. Woo, G. V. Paddock, W. Salser, J. Dorson, R. E. Moses, and B. W. O'Malley, *J. Biol. Chem., 252*:1840 (1977).
556. M.-G. Farace, M.-F. Ahellen, P.-A. Briand, C. H. Faust, P. Vassalli, and B. Mach, *Proc. Natl. Acad. Sci. USA, 73*:723 (1976).
557. E. Truelsen, K. Gausing, B. Jochimsen, P. Jorgensen, and K. A. Marcker, *Nucleic Acids Res., 6*:3061 (1979).
558. J. Weissenbach, M. Zeevi, T. Landau, and M. Revel, *Eur. J. Biochem., 98*:1 (1979).
559. C. K. Klukas, F. Cramer, and H. Gould, *Nature, 269*:262 (1977).
560. P. Little, P. Curtis, Ch. Coutelle, J. Van Den Berg, R. Dalgleish, S. Malcolm, M. Courtney, D. Westaway, and R. Williamson, *Nature, 273*:640 (1978).
561. W. L. Miller and D. C. Alexande,*J. Biol. Chem., 256*: 12628 (1981).
562. N. E. Hynes, B. Groner, A. E. Sippel, M. Chi Nguyen-Huu, and G. Schütz, *Cell, 11*:923 (1977).
563. S. H. Yu, P. P. L. Lau, and T. G. Spring, *Biochim. Biophys. Acta, 517*:31 (1978).
564. A. A. Akhundova, T. Y. Peresleni, and G. A. Dvorkin, *Biokhimiya, 43*:789 (1978).

565. B. E. Noyer, M. Mevarech, R. Stein, and K. L. Agarwal, *Proc. Natl. Acad. Sci. USA*, *76*:1770 (1979).
566. S. Z. Shapiro and J. R. Young, *J. Biol. Chem.*, *256*:1495 (1981).
567. G. M. Wahl, M. Stern, and G. R. Stark, *Proc. Natl. Acad. Sci. USA*, *76*:3683 (1979).
568. H. Kapmeyer, D. A. Lappi, and N. O. Kaplan, *Anal. Biochem.*, *99*:189 (1979).
569. A. A. Nedospasov and R. M. Khomutov, *Bioorg. Khim.*, *4*:645 (1978).
570. J. M. Rosen, S. L. C. Woo, and J. P. Comstock, *Biochemistry*, *14*:2895 (1975).
571. G. E. Sonensheim and G. Brawerman, *Biochemistry*, *16*:5445 (1977).
572. T. P. H. Tse, H. P. Morris, and J. M. Taylor, *Biochemistry*, *17*:3121 (1978).
573. I. Boime, D. Mc Williams, E. Szczesna, and M. Camel, *J. Biol. Chem.*, *251*:820 (1976).
574. D. J. Shapiro and J. J. Baker, *J. Biol. Chem.*, *252*:5244 (1977).
575. D. C. Lee, G. S. Mc Knight, and R. D. Palmiter, *J. Biol. Chem.*, *253*:3494 (1978).
576. M. A. Innis and D. L. Miller, *J. Biol. Chem.*, *252*:8469 (1977).
577. J. O. Bishop, J. G. Morton, M. Rosbash, and M. Richardson, *Nature*, *250*:199 (1974).
578. L. Chan, W. A. Bradley, and A. R. Means, *J. Biol. Chem.*, *255*:10060 (1980).
579. K. L. O'Malley, A. Mauron, J. Raese, J. D. Barchas, and L. Kedes, *Proc. Natl. Acad. Sci. USA*, *80*:2161 (1983).

Index